IMAGES OF ASIA

# The Birds of Java and Bali

# Titles in the series

At the Chinese Table
T. C. LAI

Balinese Paintings
A. A. M. DJELANTIK

The Birds of Java and Bali
DEREK HOLMES and
STEPHEN NASH

Building a Malay House
PHILLIP GIBBS

Chinese Jade
JOAN HARTMAN-GOLDSMITH

Early Maps of South-East Asia
R. T. FELL

Folk Pottery in South-East Asia
DAWN F. ROONEY

Fruits of South-East Asia:
Facts and Folklore
JACQUELINE M. PIPER

A Garden of Eden: Plant Life in
South-East Asia
WENDY VEEVERS-CARTER

The House in South-East Asia
JACQUES DUMARÇAY

Images of the Buddha in Thailand
DOROTHY H. FICKLE

Indonesian Batik: Processes, Patterns
and Places
SYLVIA FRASER-LU

The Kris: Mystic Weapon of the
Malay World (2nd ed.)
EDWARD FREY

Macau
CESAR GUILLEN-NUÑEZ

Musical Instruments of
South-East Asia
ERIC TAYLOR

Old Bangkok
MICHAEL SMITHIES

Riches of the Wild: Land Mammals
of South-East Asia
EARL OF CRANBROOK

Sailing Craft of Indonesia
ADRIAN HORRIDGE

Sarawak Crafts: Methods, Materials,
and Motifs
HEIDI MUNAN

Silverware of South-East Asia
SYLVIA FRASER-LU

Traditional Chinese Clothing
VALERY M. GARRETT

Javan Wattled Lapwing *Vanellus macropterus*

# The Birds of Java and Bali

DEREK HOLMES

*Illustrated by*
STEPHEN NASH

SINGAPORE
OXFORD UNIVERSITY PRESS
OXFORD NEW YORK
1989

*Oxford University Press*

*Oxford New York Toronto*
*Delhi Bombay Calcutta Madras Karachi*
*Petaling Jaya Singapore Hong Kong Tokyo*
*Nairobi Dar es Salaam Cape Town*
*Melbourne Auckland*
*and associated companies in*
*Berlin Ibadan*

*Oxford is a trade mark of Oxford University Press*

© *Oxford University Press Pte. Ltd. 1989*

*ISBN 0 19 588927 4*

*Printed in Singapore by Kim Hup Lee Printing Co. Pte. Ltd.*
*Published by Oxford University Press Pte. Ltd.,*
*Unit 221, Ubi Avenue 4, Singapore 1440*

# Acknowledgements

AMONG those who have encouraged or assisted with the preparation of this book, the author would first like to name W. G. (Bill) Harvey, who shared the conception of a popular book twelve years ago, while watching the birds congregating on Pulau Dua in the avian evening rush-hour, from the vantage point of the watch-tower. Paul Andrew, Bas van Balen, and S. Somadikarta have all commented on sections of the draft, and their assistance has been very valuable. Anne Nash undertook the first editing and typing of the manuscript. Several members of the Indonesian Ornithological Society have given moral encouragement through urging early completion, particularly D. Ashari, Chuck Darsono, Kamil Oesman, and Linus Simanjuntak. An Indonesian-language version is under preparation, and we are most grateful to Soenartono Adisoemarto for his willingness to undertake the difficult task of translation.

*Jakarta*            D. A. H.
*September 1988*        S. V. N.

# Contents

*Acknowledgements*   v
*Introduction*   ix

Cormorants   1
Frigate-Birds   2
Herons, Egrets, and
   Bitterns   3
Storks and Ibises   6
Ducks   8
Birds of Prey   9
Game-birds   12
Crakes and Rails   14
Plovers and Waders   17
Terns   20
Pigeons   23
Parrots   26
Cuckoos   28
Owls   31
Nightjars   34
Swifts   35
Trogons   37
Kingfishers   38
Bee-eaters   40
Hornbills   41
Barbets   42
Woodpeckers   45
Pittas   47

Larks   48
Swallows   48
Cuckoo-Shrikes   50
Minivets   52
Ioras and Leafbirds   53
Bulbuls   54
Drongos   56
Orioles   57
Crows   58
Tits and Nuthatches   60
Babblers   61
Thrushes   65
Warblers   69
Flycatchers   73
Wagtails and Pipits   75
Wood-Swallows   76
Shrikes   77
Starlings and Mynas   77
Sunbirds   80
Flowerpeckers   82
White-eyes   83
Sparrows, Weavers, and
   Munias   84

*Checklist*   88
*Index to Genera,
   Systematic Section*   106

# Introduction

IT seems deplorable that a country as rich in wildlife as Indonesia should have so few popular books on the subject. Birds alone number about 1,500 species, and up to 400 of these are endemic to the Republic, that is, they are not found anywhere else. Yet it is very difficult to obtain information about them. Comprehensive field guides that are available or under preparation for some regions may be invaluable for the serious ornithologist, but the wealth of species and technical details of their identification may only confuse or discourage the average layman who has a lively interest in his natural surroundings. The author's own interest in birds was fostered in childhood by a simple book such as this.

This book is designed to give the layman an introduction to each bird family, and by describing one or more birds from each, enable him to identify most of the birds that can be found readily in the various habitats of Java and Bali. There are colour illustrations of 112 birds, and text references to at least 120 more. A checklist at the back tabulates 433 species on the Java and Bali lists. Once the reader has developed sufficient interest to compare his own findings with this list, he will need to obtain a more detailed field guide. The list is not fully comprehensive, as there are some 50 additional species which reach the islands only rarely, as migrants off-course or as oceanic wanderers, and these are not included.

Two factors contribute to Indonesia's wealth of birds. One is the tropical rain forest, which is the richest environment in the world in terms of species diversity. The second is Indonesia's unique position, spanning two geological continental areas. The western islands of Sumatra, Kalimantan, and Java (with Bali) lie on the Sunda Shelf, linking them with the Malay Peninsula and the Asian continent, while the island of Irian Jaya, with Papua New Guinea, lies on the Sahul Shelf, linking it with the Australian continent. During the successive lowering of sea levels in the Ice Ages, these continental shelves were exposed as dry land, permitting free interchange of fauna. The remaining islands of Sulawesi, Maluku, and Nusa Tenggara are surrounded by deep seas; they constitute the region known as 'Wallacea', which has faunas that are transitional between the two continental plates, as well as a large number of endemic species.

Thus Java and Bali lie at the southern extremity of the old Sundanese continent, but they have been isolated for sufficient time to allow new species to evolve. The two islands, with a gross land area of 138 000 sq. km, may have fewer species than the larger neighbouring island of Sumatra (gross area of 475 000 sq. km, with its satellite islands), but they have up to 31 endemic species, which is about double the number found on Sumatra. Furthermore, the drier and more monsoonal climate of East Java and Bali permits a wider range of open-country species.

The checklist at the back indicates over 340 resident birds on Java; the remainder are migrants or nomadic. It will be seen that many occur on Java but not Bali. This is characteristic of smaller islands, and Bali, which is 5 600 sq. km in area, has only about half this number of resident birds. Little is known about the birds of Madura, but it is likely that its avifauna has become very impoverished, as a result of the limited variety of habitats that now exist on that island. Smaller islands, such as Nusa Penida off Bali, the Thousand Islands off Jakarta, and the Karimun Jawa, Bawean, and Kangean Islands further offshore, have a much reduced avifauna, but include some small island specialists and species originating from adjacent regions which are not discussed in this book.

Visitors often comment that Java has very few birds. Alas, in many areas, this is true. A human population of over 100 million with its land requirements for food production does not leave room for many wild habitats. However, the diligent naturalist can find good localities, and those who have the time to visit the national parks and reserves, such as Ujung Kulon, Gunung Halimun, Gunung Pangrango-Gede, Meru-Betiri, Baluran, and Bali Barat, will experience a taste of the region's former abundance of birds. Even close to the two biggest cities of Jakarta and Surabaya, there are large colonies of water-birds breeding in the mangroves (Pulau Dua, Pulau Rambut, and the river deltas of East Java).

Nevertheless, the loss of forests and the decline in bird populations is a source of major concern. Huge areas of forest on Java and Bali were already lost long before scientific studies even commenced, and open-country species continue to be decimated by trapping, farm chemicals, and airguns. A major objective of a popular book on natural history is to increase awareness of the need to protect and conserve wildlife for the future.

To the author's knowledge, only one endemic bird in Java has become extinct. This is the Javan Wattled Lapwing, a plover that must have been a striking bird of coastal grasslands (see Frontispiece). It has not been seen for over three decades. There is, however, a large number of birds that have become very rare in Java and Bali, and they could soon become extinct unless urgent action is taken to protect them and their remnant habitats.

The colour plates illustrate principally the male birds. No scales are given, as individual plates cover a rather wide range of species, but sizes are given in the text. The novice bird-watcher should beware of variations between subspecies that occur in different areas of Indonesia. For example, on Plate 12, the Coppersmith and Blue-eared Barbets occurring in Sumatra have a different distribution of head colours. In a few cases, such as the Magpie Robin, there are quite marked variations even within Java and Bali.

Both English and scientific names are given, and those with more than a passing interest are advised to become familiar with the latter, as English names may vary from book to book. Indonesian names are given for many, sometimes only a generic name for the group, but it should be remembered that different names may be used among the various languages and dialects. There is still no common Indonesian name for each of the 1,500 species found in the Republic, although names are being compiled.

# Cormorants

CORMORANTS are black, long-necked water-birds that feed on fish by diving from the surface. In tropical regions they roost mainly in trees. They are gregarious and often seen in quite large flocks. When resting, they have the habit of holding out their wings to dry, looking like so many witches.

## LITTLE CORMORANT
*Phalacrocorax niger* (50 cm)
**Pecuk**                                                      Figure 1

There are two species of cormorant on Java, but most bird-watchers have difficulty in distinguishing between them. The Little Cormorant, which has an oriental distribution, is common in Java though it is not often found elsewhere in

Indonesia. However, Java is also the western outpost of an Australian species, the LITTLE BLACK CORMORANT *Phalacrocorax sulcirostris*.

Cormorants breed colonially in the mangroves of Pulau Rambut, Pulau Dua, and the Solo delta region. Up to 1,200 pairs breed on Pulau Dua. In Jakarta, small groups can often be seen at Ancol, and sometimes hundreds perch on the electric cables by the road to Soekarno–Hatta Airport. The evening flights homing in to Pulau Dua from inland, or along the coast, make an impressive sight.

1. Little Cormorant,
   *Phalacrocorax niger*

Cormorants should not be mistaken for the rarer ORIENTAL DARTER *Anhinga melanogaster* (90 cm), which occurs in similar habitats, and whose extremely long, sinuous neck gives it the popular name of 'Snakebird'. A few darters are always present by the road to Soekarno–Hatta Airport and at Pulau Dua, and some still breed at Pulau Rambut. There are no recent records from Bali.

# Frigate-Birds

THE ordinary, land-based bird-watcher in Java is not likely to see many sea-birds, except for frigate-birds which often soar just off-shore, sometimes even at Jakarta. Graceful against the wind, they have a very distinctive silhouette with their long bodies and forked tails, and very long, pointed wings angled in the middle section. They are mainly black, with varying amounts of white on the under-parts according to species, sex,

2. Lesser Frigate-Bird *Fregata ariel*

and age. They feed by pirating upon other sea-birds, chasing them until they drop their prey, but they are not averse to pick offal thrown overboard, when they look more like a gaggle of witches than masters of the air currents. They breed on rocky islands, where the males acquire an inflatable red pouch below the bill. That illustrated in Figure 2 is the adult male LESSER FRIGATE-BIRD *Fregata ariel* (80 cm), the species most often encountered in Java and Bali.

# Herons, Egrets, and Bitterns

HERONS are tall wading birds with long or very long necks. They feed by walking through shallow waters and marshes, stabbing at fish and crustaceans with their dagger-like bills. In flight, their rounded wings beat steadily, and the neck is drawn back in an 'S'-shape (unlike storks, which fly with the neck extended). Most egrets are white and are very conspicuous when they feed in open swamps, ricefields, and pastures. By contrast, the bitterns hide in reed-beds or dense swamp vegetation and have cryptic colours. _

GREY HERON
*Ardea cinerea* (100 cm)
**Cangak Abu**                                                      Plate 1

The Grey Heron is a pale grey bird with a white head and neck, and a black mask behind the eye that leads back to form a fine crest. In the oriental region, it is mainly a bird of coastal areas, whereas the slightly smaller PURPLE HERON *Ardea purpurea*, which is dark rufous in colour, feeds in .fresh water. Both birds may be seen at Pulau Dua, though only the Grey Heron now breeds there, in small numbers. Both species range from Africa and Europe through Asia to the western parts of Indonesia.

## JAVAN POND HERON
*Ardeola speciosa* (45 cm)
**Blekok Sawah** Plate 1

This is the small brown and white heron that can be seen in the ricefields, both on the coast and in the hills. When settled, it appears dull brown and may escape detection, but its pure white wings are startling when it takes flight. During the breeding season, the back becomes blackish and the breast is cinnamon. Over 900 pairs breed on Pulau Dua between February and June. This Pond Heron's range extends east to Sulawesi and Sumba, but very closely related species are found widely in India and the oriental region.

## CATTLE EGRET
*Bubulcus ibis* (50 cm)
**Kuntul Kerbau** Plate 1

The Cattle Egret, the smallest and commonest of the egrets, is so named because of its partiality for feeding near grazing cattle. Its build is more graceful than the Pond Heron, though not as slender as the larger egrets. Its plumage is all white, but in the breeding season, the head, neck, and back acquire buff plumes. The bill is yellow, and thicker than in other egrets. This is the most abundant breeding bird on Pulau Dua, with 3,500 pairs, and its range extends from India to Sulawesi and Nusa Tenggara.

## LITTLE EGRET
*Egretta garzetta* (60 cm)
**Kuntul Kecil** Figure 3

The Little Egret is larger and more slender than the Cattle Egret, and is also distinguished by its fine black bill and legs. During the breeding season, it acquires two long, narrow white plumes on the nape, and further plumes on the back which extend beyond the tail. Formerly, these delicate feathery plumes were much in demand for adorning the hats of fashionable ladies, but the birds are now protected. Large flocks congregate to roost in the mangroves at Pulau Rambut and Pulau Dua, with over 1,000 pairs breeding on Pulau Dua.

However, care must be taken to distinguish them from the larger PLUMED EGRET *Egretta intermedia* (70 cm) and the GREAT EGRET *Egretta alba* (90 cm).

There is one other egret at Pulau Dua, the PACIFIC REEF EGRET *Egretta sacra*, which is more or less confined to the littoral zone. The same size as the Little Egret, it occurs in two distinct colour phases: the dark phase is easy to recognize, being slate-coloured, but the white phase presents some problems (look for the greenish legs and flesh-coloured bill).

3. Little Egret *Egretta garzetta*

## BLACK-CROWNED NIGHT HERON
*Nycticorax nycticorax* (60 cm)
**Kowak**                                                    Plate 1

Among the black cormorants and white egrets at Pulau Dua are more squat herons, blue–black above and white below, whose rather stubby appearance recalls the Pond Herons. Unlike other birds that fly off at dawn and return at dusk, the Night Herons feed mainly by night, though they may be active throughout when feeding chicks. Quite often their harsh 'kwark' may be heard from birds flying over Jakarta at night from Pulau Rambut or the marshes at Cengkareng. In recent years, a large roosting colony has formed on the island of the ornamental lake in the Bogor Botanical Gardens. In breeding plumage, the Night Heron also has two long white plumes extending down the mantle. Immature birds have dull brown plumage with bold whitish spots.

## CINNAMON BITTERN
*Ixobrychus cinnamomeus* (40 cm)        Plate 1

The bitterns are small members of the heron family that spend most of their time hidden in reed-beds and tall swamp grass. They are most likely to be seen when they make short flights over the swamp to seek a new feeding area at dawn or to return to their roosting site at dusk. Unlike other herons, they do not roost in trees but stay on, or near, the ground. The Cinnamon Bittern is quite a common bird in Java, and can often be seen flying over ricefields, marshes, or scrubby damp thickets, and even in the hills. The drawn-in neck and rounded wings of uniform chestnut colour are distinctive. The under-parts are paler and streaked, especially down the centre of the neck and breast. When disturbed, bitterns will attempt to escape detection by immobilizing themselves, with neck outstretched almost vertically, but it is rare that one can creep close enough to observe this behaviour. This species occurs widely from India to China, south to Java, Sulawesi, and Flores.

# Storks and Ibises

STORKS are tall, gaunt birds with long legs and necks, and massive, dagger-like bills. Ibises have a somewhat less elongated appearance, except for their long bills which are fine and strongly down-curved. Unlike herons, they fly with neck extended, and storks in particular are expert at soaring in thermals. They feed on mud-flats and open ground, and roost and breed in trees, usually colonially. In many parts of the world, these large and striking wading birds are endangered, and the five species in Java—three storks and two ibises—have now all become rare.

## WOOLLY-NECKED STORK
*Ciconia episcopus* (90 cm)
**Sandang Lawe**                                                  Plate 2

In East Java, this stork is perhaps the least rare of the three, although its breeding sites are not yet known. Generally favouring damp areas, it will also feed on pasture land or fallows, in parties of up to a dozen. It is easily identified by its black body and wings, and white neck and belly. The legs are red. Its range extends from Africa and India, through Sumatra and Java, to Sulawesi and the Philippines, though it is absent from Kalimantan.

The biggest stork is the LESSER ADJUTANT *Leptoptilos javanicus* (115 cm), an ugly, evil-looking bird of coastal areas, mainly black but with white under-parts and a massive bill.

The third species is the MILKY STORK *Mycteria cinerea* (100 cm), a more graceful bird with a slightly down-curved bill, predominantly white except for a dark breast band and black tail and flight feathers. This stork, which occurs from Kampuchea down to Sumatra, Java, and Sulawesi, is an endangered species, but recently a population of over 3,000 has been discovered along the east coast of Sumatra. Up to 14 pairs breed on Pulau Rambut, but it has abandoned Pulau Dua as a breeding site since the 1970s. A few can sometimes be seen along the road to Soekarno–Hatta Airport; there are few international airports in the world where the visitor has the chance to see an endangered species immediately on arrival!

## GLOSSY IBIS
*Plegadis falcinellus* (65 cm)
**Roko-roko**                                    Plate 2

When feeding, ibises are more active and less sentinel-like than storks. The Glossy Ibis is generally gregarious, feeding in small parties in marshes and sometimes in ricefields, in Java and Sulawesi. It appears all black but, in a good light, it is actually a glossy dark chestnut. Flocks of 50 or 60 are occasionally encountered around Surabaya, and about 40 pairs breed at Pulau Dua, where there may also be a few roosting BLACK-HEADED IBIS *Threskiornis melanocephalus*. This is a larger ibis (75 cm), mainly white but with a black head and bill; it has a substantial population on the east coast of Sumatra.

# Ducks

## LESSER TREEDUCK
*Dendrocygna javanica* (40 cm)
**Belibis**                                      Plate 2

The only common duck in Java is the Lesser Treeduck, perhaps better known as the Whistling Teal. Its Indonesian name aptly describes its mellow whistling calls, frequently uttered in flight. It occurs on lakes, ponds, and flooded ricefields, even around the outskirts of Jakarta. It is brown all over, of various shades, with a dark cap and chestnut on the wings. It is found from India to western Indonesia.

A larger species, the WANDERING TREEDUCK *Dendrocygna arcuata* (48 cm), appears to replace it in East Java and Bali. This is an eastern species, with a range extending to the Philippines and northern Australia. It is very similar to the Lesser Treeduck, but has pronounced white patches on the flanks and rump.

The other duck that is seen regularly is the GREY TEAL

*Anas gibberifrons* (43 cm), a bird inhabiting coastal areas and mangroves. It is an indistinctly plumaged true duck, mottled greyish-brown, paler below and about the head. In flight, there is white at the base of the under wing, and white, black, and green in the upper wing. It is mainly seen in twos and threes, and there are usually a few beside the road to Soekarno–Hatta Airport.

# Birds of Prey

THE birds of prey (sometimes known as raptors) form a wide group of predatory birds that seek their prey from the air by a variety of methods. Most of the kites, buzzards, and eagles soar, dropping on the prey feet first. Sparrowhawks have swift darting flights through the tree-tops, and falcons dive at high speed. Some species hover. Body size and shape vary widely to accommodate these different hunting methods.

## BLACK-SHOULDERED KITE
*Elanus caeruleus* (33 cm)                                Plate 3

Only two raptors in Java regularly hover suspended in mid-air while hunting for their prey on the ground–the Black-shouldered Kite and the Spotted Kestrel (see p. 12). The Black-shouldered Kite is a rather small raptor of the open country-side, notable for its pale plumage and pointed wings. Its upper-parts are pale grey, with a black patch on its shoulders, while its under-parts are whitish. Its tail is not forked as in true kites. This bird has a graceful, gliding flight punctuated by short spells of hovering, prior to plunging to the ground to catch mice, grasshoppers, and other small prey. It has a wide distribution, from Africa and southern Europe to China and Papua New Guinea. In Indonesia, it occurs on all the islands except Maluku and Irian Jaya, but it has now become scarce in Java.

# WHITE-BELLIED SEA EAGLE
*Haliaeetus leucogaster* (70 cm)
**Elang Laut** Plate 3

This magnificent eagle can be seen locally along the coasts, usually in pairs. Two pairs are said to breed on Pulau Rambut. The bird chooses a prominent perch in an isolated tall tree near the beach for resting, and soars high over the sea close to shore. It is mainly white below, with broad wings and a short, wedge-shaped tail, but the flight feathers and base of the tail are black. The upper-parts are grey. Immature birds do not have the white under-parts. The call is loud, nasal, and shrill. The species occurs throughout Indonesia, and from India to Australia.

The White-bellied Sea Eagle should not be mistaken for the BRAHMINY KITE *Haliastur indus* (45 cm), a much smaller raptor of coastal areas and rivers, often near towns and fishing ports. This has a distinctive chestnut plumage, but with white head, throat, and breast. Sadly, although such a familiar bird of coasts all the way from India to northern Australia, it has become very scarce in Java.

# CRESTED SERPENT-EAGLE
*Spilornis cheela* (60 cm) Plate 3

There are several raptors of the eagle type in Java which are difficult for the novice to identify. They all have long, broad wings, fingered at the tip, and spend much time soaring. The Crested Serpent-Eagle is quite common in wooded areas of the lowlands and lower hills, and is more easily identified from below than most, having one broad white band in the tail and another white band close to the hind margin of the under wing. When perched, it seems dark brown, spotted on the under-parts, and with a pronounced crest. The call is a shrill di- or tri-syllabic note which often draws attention to a soaring bird or pair. It is common throughout the Sundanese region.

A larger, all-black eagle seen soaring in similar terrain is likely to be the BLACK EAGLE *Ictinaetus malayensis* (70 cm). Some care is needed in identification, however, as the CHANGEABLE HAWK-EAGLE *Spizaetus cirrhatus* (70 cm) occurs in various colour phases, as its name implies, and a few can be nearly as black. Particularly interesting is the JAVAN HAWK-EAGLE *Spizaetus bartelsi* (60 cm), which is endemic to Java. It is a broad-winged, rather long-tailed eagle, with a long erectile crest. It is brown above but paler below, with dark stripes on the breast and strong barring on the belly. It has three broad, dark bands in the tail. The crest is present even in the young bird, which is tawny-coloured. It may have become one of the world's rarest birds, although it can still be seen quite readily over wooded areas of Gunung Gede, Gunung Pangrango, Gunung Halimun, and Meru-Betiri.

## JAPANESE SPARROWHAWK
*Accipiter gularis* (28 cm)                    Plate 3

The sparrowhawks have short, broad wings and rather long tails, and hunt by rapid flight at or below tree-top level, preying mainly on small birds. Even larger pigeons fly off in alarm when one passes overhead. The different species are not easy to identify, but the Japanese Sparrowhawk is a common migrant during the northern winter, and is the species most often encountered. It is grey (male) or brown (female) above, and barred pale brown and white below.

The migrant hawks often move in loose flocks, sometimes numbering hundreds, south in October and north in March–April. These flocks have been well documented in the Malay Peninsula, but much less is known of their continuing passage through Sumatra and then east through Java, Bali, and beyond. Several species are involved in these migrations, though the most numerous are the Japanese Sparrowhawk and the larger HONEY BUZZARD *Pernis apivorus* (50 cm).

## SPOTTED KESTREL
*Falco moluccensis* (32 cm)
**Alap-alap** Plate 3

Of the three falcons resident in Java, the Spotted Kestrel is the commonest. Falcons have long, narrow wings and tail, and swift, dashing flight, but the Kestrel regularly hovers in one spot on beating wings for prolonged periods, with short flights to take up new hovering positions. It catches its prey by dropping to the ground with outstretched talons. The Kestrel is seen widely in open country in the lowlands of Java, Sulawesi, Maluku, and Nusa Tenggara. It is strongly rufous in colour, spotted black. The head is brownish with a pronounced dark 'moustache' in the male. The tail is bluish-grey with a black band near the white tip. The call is a sharp, rapid 'kee-kee-kee-kee-kee'.

# Game-birds

THE game-birds include the ground-dwelling quails, partridges, and pheasants, which are everywhere favourites for the pot. Several species have become domesticated and are farmed. In Java, true pheasants are absent, and as a group, game-birds are not well represented.

## BLUE-BREASTED QUAIL
*Coturnix chinensis* (15 cm)
**Puyuh** Plate 4

There are three kinds of small quail in Java, all difficult to identify as usually they are seen merely as a small bundle of rufous brown whirring away from the observer and diving into cover. They live in grassland, dryland crops and scrub, and run through the ground cover in order to remain hidden. The short whirring flight is insufficient to establish identity, so a combination of stealth and chance is required to obtain a good view. The male is distinctively marked, as shown in Plate 4, and the call is also distinctive—a quiet, plaintive 'pi-pi-piu'.

However, the female is dull brownish in colour, and very similar to the slightly larger BARRED BUTTON-QUAIL *Turnix suscitator*, the female of which has a black throat. There is also a smaller species, the LITTLE BUTTON-QUAIL *Turnix sylvatica*, so these birds present quite a problem to the casual bird-watcher. Probably the Barred Button-Quail is the commonest of the three. It is polyandrous, with the males attending to incubation, and the female advertises herself with soft but far-carrying booming and purring notes.

## GREEN JUNGLEFOWL
*Gallus varius* (male 60 cm, female 42 cm)
**Ayam Hutan Hijau**                                    Plate 4

The Green Junglefowl is endemic from Java east to Flores and Sumba. Another species, the slightly larger RED JUNGLE-FOWL *Gallus gallus*, which is the ancestor of the domestic chicken, has a much wider distribution, from the Himalayas through South-East Asia to Sumatra, Java, Sulawesi (where it was introduced), Lombok, and Timor. Both junglefowl occur at all altitudes in Java, but the Red prefers the wilder margins of primary forest, while the Green tolerates more open terrain. The author found the Green Junglefowl very common in the coffee plantations on the Ijen Plateau in East Java, but the following week was surprised to hear it on the hot, rocky, and scrubby hillsides of the south coast. The calls are quite distinct, the Red's being somewhat similar to that of the familiar domestic cockerel, while the Green's is a thinner, more nasal, three-note call. The Green Junglefowl is a dark bird, but in a good light has a beautiful iridescent green sheen; the comb is violet and red. The Red Junglefowl has the typical red comb and wattles of the domestic bird.

Visitors to Ujung Kulon and Baluran will become aware of the huge GREEN PEAFOWL *Pavo muticus*, well known in zoos and famous for its brilliant long tail plumes and loud wailing cries. Perhaps it is inevitable that such a beautiful bird should become endangered in Java, not least because its tail plumes are used for ceremonial purposes, and while it was formerly quite widely distributed, it is now restricted to the two national parks at either end of the island.

The endemic JAVAN PARTRIDGE *Arborophila javanica* (27 cm) is quite common in the mountain forests of West and Central Java. It occurs in small parties that are usually seen as dark, rounded birds running off the trail or whirring away through the trees. If the observer manages to get a good view, it is seen to have a strongly patterned head of chestnut, rusty red, and black. The upper-parts are grey, barred with black; the breast is grey and the belly chestnut. The legs are red. The most striking feature is the call—a series of whistled notes uttered in duet between male and female, increasing in intensity but never reaching a climax. In East Java, it is replaced by a race of a more widespread oriental species, the BAR-BACKED PARTRIDGE *Arborophila orientalis*, which has white stripes on the head, and black and white barring on the flanks.

# Crakes and Rails

THIS group consists of marsh birds that swim or wade in reedbeds and adjacent mud. They fly weakly, with dangling legs, and usually only for very short distances, though some are capable of long migratory flights. They have short, rounded wings and short tails that are often cocked to reveal a rufous, white, or barred patch below. The larger coots and moorhens are quite conspicuous but the smaller rails and crakes are extremely secretive and difficult to observe.

## WHITE-BREASTED WATERHEN
*Amaurornis phoenicurus* (33 cm)
**Kareo**                                                    Plate 4

This conspicuous bird frequents roadside ditches, damp thickets, and ricefields. It readily comes out into the open, though never walking far from cover, and it will quickly run or fly back when disturbed. It is easy to identify: a rather slender bird with a very short tail and moderately long bill and legs. It is dark greyish-brown but with a prominent white face, throat, and breast. Its tail is commonly cocked up when it walks, revealing chestnut under tail coverts. However, it is best known for its extraordinary call—a real cacophony—beginning with grunts and chuckles and then developing into a prolonged series of 'kru-wak, kru-wak' notes, sometimes for minutes on end. It occurs from India and China to the Philippines, Sulawesi, and Nusa Tenggara, and in Java it follows the ricefields high into the mountains.

## PURPLE SWAMPHEN
*Porphyrio porphyrio* (43 cm)
**Mandar Besar**                                             Plate 4

In the larger swamps and reed-beds of the lowlands, one may find the Purple Swamphen, a heavy bird of bluish-purple colour, with red bill and legs, and white under the tail. Generally it remains hidden in the reed-beds, and stealth is required to obtain a view, except when it makes brief flights, with dangling legs, over the top of the vegetation.

This bird has a very wide distribution, from Africa to New Zealand and some islands in the Pacific. However, an observer is more likely to get good views of another widely distributed bird, the COMMON MOORHEN *Gallinula chloropus* (33 cm). The moorhen occurs in smaller patches of swamp and will often swim out of cover. It is dark slaty to brownish, but with two oval white patches beneath the tail, a white line along the flanks, and a yellow-tipped red bill with a red knob at its base. It has a characteristic sharp, liquid 'prrrt' call. It occurs throughout western Indonesia with a closely related Australasian species, the DUSKY MOORHEN *Gallinula tenebrosa*, replacing it further east.

## WHITE-BROWED CRAKE
*Porzana cinerea* (20 cm)
**Tikusan Alis Putih**                                      Plate 4

The White-browed Crake is the most readily seen of the smaller rails and crakes, as it often comes into the open areas adjacent to the reeds and rushes of the small marshes that it frequents. Brown above, grading to whitish on the under-parts and buff under the tail, its diagnostic feature is the grey head and neck with two prominent white streaks on the face, above and below the eye. The bill is reddish and the legs greenish. It is quite vocal, with shrill, piping 'weeu, weeu, weeu' calls, especially early and late in the day. It is found from South-East Asia to Australia.

There are several species of rail but they are shy and difficult to see adequately. In swamps like those along the road to Soekarno–Hatta Airport, one is likely to see both the RUDDY CRAKE *Porzana fusca* (22 cm), which is dark rufous brown all over with some barring on the belly, and the SLATY-BREASTED RAIL *Rallus striatus* (25 cm), with a chestnut cap, grey face and breast, and heavy white barring across the body. The latter bird can also be flushed sometimes from dry *alang-alang* grasslands.

All the rails are interesting to study, and the best method is to position oneself quietly at dawn or dusk on bunds over-looking the edges of swamp vegetation and wait for them to stealthily creep from cover. Any sharp movement or noise will send them scurrying back to shelter.

# Plovers and Waders

GROUPED here are the numerous wading birds that are generally found near water, especially on mud-flats. They range in size from the stints (15 cm) to the tall greenshanks (35 cm) and long-billed curlews (58 cm). Most species are migratory, visiting Java during the northern winter. They are generally brown, and even experts have difficulty in identifying the smaller species. The true waders, which include the sandpipers, godwits, curlews, and snipe, have fine legs and bills, while the plovers are more squat in build, with shorter bills. Only a few characteristic species are described here.

GOLDEN PLOVER
*Pluvialis dominica* (25 cm)
**Trulek Kli-it** Plate 5

Among the many migrant waders that visit Indonesia during the northern winter is the Golden Plover, a rather plump bird that especially favours short grass. From October to April, residents near golf-courses, football pitches, and airfields will often hear their whistled 'tu-ee' calls as they fly in at night to roost. The plumage is brown, mottled at close range, with pale under-parts and uniform, unpatterned wings. The short bill is dark. The breeding plumage is quite different, with golden mottles and black under-parts, and some birds, either early or late in the wintering season, may be in moult, showing dark blotches on the belly.

## JAVAN WATTLED LAPWING
*Vanellus macropterus* (28 cm)                     Frontispiece

It is at least three decades since this endemic plover has been seen, and it must now be considered extinct. As the frontispiece illustrates, it was a striking bird, particularly when the white under wings show in flight (see title page), but it was also conspicuous and perhaps noisy, making it an early candidate to succumb to the pressures of heavy human population and hunting. Its habitat was said to have been the wet grassland around inland swamps or the backs of coastal dunes.

## LITTLE RINGED PLOVER
*Charadrius dubius* (18 cm)
**Cerek Kalung Hitam**                              Plate 5

The several plovers of this genus are not easy to distinguish, but the migrant Little Ringed Plover is more distinctly marked, and the beginner should not have difficulty in identification. Plate 5 shows a bird in winter plumage, or an immature, typical of most seen in Indonesia, though occasionally one finds an adult in breeding plumage, in which the breast and collar band is black, and there is a black mask through the eye and across the forecrown. Note that the breast band completely encircles the body, separated from the back of the head by a white collar. In flight, the outer tail feathers are seen to be white, but the uniform upper wing, without a pale wing bar, is distinctive of this species. Further characteristics are the yellowish legs and the 'pee-u' call. Favouring mud-flats and sands on or near the coast or rivers, it likes the partially dry brackish-water fish-ponds of Java's north coast.

The JAVAN SAND PLOVER *Charadrius javanicus* is a small resident plover of sandy beaches and adjacent areas. Sandy brown above and white below, it differs from the Little Ringed Plover in having a narrow white wing bar, and in lacking a complete black breast band, which is reduced to black patches on the sides of the breast. It is endemic to Java, although some authors treat it as a race of the KENTISH PLOVER *Charadrius alexandrinus*, which is a winter migrant. In Bali and the other islands of western Indonesia, it is replaced by the very similar MALAY SAND PLOVER *Charadrius peronii*. There are, however, several species of small plovers and waders that visit Java in winter, and most of them have white wing bars, so they are difficult to identify.

## WHIMBREL
*Numenius phaeopus* (43 cm)
**Gajahan**                                              Plate 5

While the identification of the numerous small waders of the coastal mud-flats during the northern winter is beyond the scope of this book, the larger curlews and whimbrels are distinctive because of their very long, curved bills. The Whimbrel favours mud-flats close to mangroves and will even wander into the mangroves on occasion. Brown above and pale below, it has a whitish triangle extending up its back, prominent in flight. The larger EURASIAN CURLEW *Numenius arquata* (58 cm) has an even longer bill than the Whimbrel. To distinguish the two species, listen for the Whimbrel's rapid piping 'pi-pi-pi-pi', distinct from the Curlew's shrill, rising 'cour-lee'.

## COMMON SANDPIPER
*Actitis hypoleucos* (20 cm)
**Trinil Pantai**                                        Plate 5

Another migrant wader which is easy to identify is the Common Sandpiper. Widespread near water, including harbours and large inland rivers, it is not averse to perching on rocks or concrete piers, or resting on floating logs. It constantly bobs its body up and down when feeding. In its short, jerky flights, usually accompanied by a shrill 'twee-wee-wee', it alternates rapid wing-beats with short glides on down-curved wings, showing its prominent white wing bar and white outer tail feathers. It is seen mostly from September to April.

Quite common in small parties in ricefields is the WOOD SANDPIPER *Tringa glareola* (23 cm). It is completely dark above, except for the prominent white upper tail coverts seen in flight. The under-parts are whitish, but the wings are all dark without a white bar. Its softer, more plaintive 'tiss-iss-iss' call often rings across the ricefields during the northern winter.

# Terns

VISITORS to Indonesia from temperate regions express surprise at the lack of sea-birds around the coasts, as gulls very rarely reach these latitudes, and terns are not common in most areas, or they stay out at sea. Terns are slender, narrow-winged sea-birds with long, pointed bills and long, forked tails; most of them are white.

## GREAT CRESTED TERN
*Sterna bergii* (46 cm)
**Dara Laut Jambul Besar**                                     Figure 4

The Great Crested, the largest tern in Indonesia, roosts along the coast on sandbanks and fishing stakes, or on floating logs out at sea. It is mostly white, with a black crest but white forecrown. The bills of many terns are yellow to reddish, but the massive, greenish-yellow bill of this species is distinctive. These terns feed by diving for fish from a medium height.

It is a partial resident, with numbers augmented from the north in winter, along with a smaller species that is locally common, the COMMON TERN *Sterna hirundo* (34 cm).

1. (a) Cattle Egret *Bubulcus ibis*. (b) Grey Heron *Ardea cinerea*. (c) Javan Pond Heron *Ardeola speciosa*. (d) Black-crowned Night Heron *Nycticorax nycticorax*. (e) Cinnamon Bittern *Ixobrychus cinnamomeus*.

2. (a) Woolly-necked Stork *Ciconia episcopus*. (b) Glossy Ibis *Plegadis falcinellus*. (c) Lesser Treeduck *Dendrocygna javanica*.

3. (a) Black-shouldered Kite *Elanus caeruleus*. (b) White-bellied Sea Eagle *Haliaeetus leucogaster*. (c) Spotted Kestrel *Falco moluccensis*. (d) Japanese Sparrowhawk *Accipiter gularis*. (e) Crested Serpent-Eagle *Spilornis cheela*.

4. (a) Green Junglefowl *Gallus varius*. (b) White-browed Crake *Porzana cinerea*. (c) Blue-breasted Quail *Coturnix chinensis*. (d) Purple Swamphen *Porphyrio porphyrio*. (e) White-breasted Waterhen *Amaurornis phoenicurus*.

5. (a) Whimbrel *Numenius phaeopus*. (b) Common Sandpiper *Actitis hypoleucos*. (c) Golden Plover *Pluvialis dominica*. (d) Little Ringed Plover *Charadrius dubius*.

6. (a) Black-naped Fruit Dove *Ptilinopus melanospila*. (b) Spotted-necked Dove *Streptopelia chinensis*. (c) Grey-headed Green Pigeon *Treron griseicauda*. (d) Little Cuckoo-Dove *Macropygia ruficeps*.

7. (a) Javan Turtle-Dove *Streptopelia bitorquata*. (b) Red-breasted Parakeet *Psittacula alexandri*. (c) Javan Hanging Parrot *Loriculus pusillus*. (d) Plaintive Cuckoo *Cacomantis merulinus*. (e) Zebra Dove *Geopelia striata*.

8. (a) Collared Scops Owl *Otus bakkamoena*. (b) Barn Owl *Tyto alba*.
(c) Chestnut-breasted Malkoha *Phaenicophaeus curvirostris*. (d) Lesser
Coucal *Centropus bengalensis*.

4. Great Crested Tern *Sterna bergii*

Mainly whitish, this latter species has a darker mantle and upper wing and, in winter especially, a dark band along the leading edge of the inner wing. The bill is black but may be reddish at the base.

Much smaller white terns around the coast are likely to be the LITTLE TERN *Sterna albifrons* (25 cm), with a black cap in breeding plumage, but a white forehead, and a yellow bill. The flight is rather jerky. Off-shore, terns with very dark–almost black–upper-parts, but with a white leading edge to the wing and a white eye-stripe, are likely to be the BRIDLED TERN *Sterna anaethetus* (37 cm). They have a deeply forked tail and gliding flight.

## WHITE-WINGED TERN
*Chlidonias leucopterus* (25 cm)
**Dara Laut Sayap Putih**                                    Figure 5

Terns of this genus, known as marsh terns, favour fresh- or brackish-water ponds and shallow muddy tidal waters. They feed by quartering restlessly back and forth at low height and plucking insects from the surface. Commonly seen over rice-fields, they are less graceful and slender than sea-terns, and

5. White-winged Tern *Chlidonias leucopterus*

their tails are only slightly forked. In breeding plumage, the head and body are black, and the rump, tail, and upper wings are white, but most birds in Indonesian waters are in their winter plumage, mainly white and grey with a black patch on the rear of the crown and ear coverts.

Winter visitors from the north, they are sometimes very common along the north coast at times of migration, when several hundreds may be feeding in the marshes near the Soekarno–Hatta Airport, or just off-shore at Muara Angke. However, there is another species, the WHISKERED TERN *Chlidonias hybridus*, which is mainly a visitor from Australia (although a few also arrive from the north). During the migration seasons, the two species often overlap, and in winter plumage they are very alike. Usually, a few specimens can be found in breeding plumage, in which the Whiskered Tern has prominently contrasting white cheeks and blacker wings. Whiskered Terns generally predominate from May to September, and the more numerous White-winged Terns from October to April, but the periods of overlap are quite prolonged.

# Pigeons

PIGEONS and doves are familiar to many people from their beautiful but sometimes mournful cooing, and some of them are popular as cage-birds. They build the flimsiest of nests which often consist only of a few sticks, on which they lay one or two white eggs. The family includes several groups or genera which differ widely, some of them being beautifully coloured.

GREY-HEADED GREEN PIGEON
*Treron griseicauda* (27 cm)
**Punai Manten**                                    Plate 6

As implied by their generic name, these pigeons are predominantly green, though the males of some species have a maroon mantle and wing coverts. One such species is the Grey-headed, which is green except for a greyish head, yellow bars in the flight feathers, chestnut under tail coverts, and red feet. The female lacks the maroon but is otherwise similar. They are exclusively arboreal, gathering in small flocks that feed on fruits in the canopy, sometimes joining other pigeons, barbets, and hornbills. This species is widespread in Java and Bali, but is common only in forests and well-wooded country. Parties may be seen swooping through the trees at dusk in the Botanical Gardens at Bogor.

In open country, one is more likely to see the PINK-NECKED GREEN PIGEON *Treron vernans*, the male of which lacks a maroon back. Instead, it has a pinkish neck and upper breast and an orange band across the lower breast.

In forests, look for the larger imperial pigeons. The GREEN

IMPERIAL PIGEON *Ducula aenea* (43 cm) of the lowlands is greyish with bronzy green mantle and wings, and dark chestnut under the tail. It has a mournful cooing call, though at close range, a clicking sound can be heard to precede it. In the mountains, it is replaced by the DARK-BACKED IMPERIAL PIGEON *Ducula lacernulata*.

## BLACK-NAPED FRUIT DOVE
*Ptilinopus melanospila* (23 cm)
**Walik**                                                      Plate 6

The fruit doves are the most colourful of the family. This species occurs in wooded areas of the lowlands and hills, and may be seen in small parties in the Bogor Botanical Gardens. The body is green, but the male is very distinctive as Plate 6 shows. The female is mainly green, with a yellow eye. The call is a slowly repeated, soft 'to-whooo'. It is now rare in Bali.

At higher elevations, the Black-naped is replaced by the PINK-NECKED FRUIT DOVE *Ptilinopus porphyreus*, a beautiful bird with deep pink head, neck, and upper breast bordered by narrow white and black bands across the breast. Its belly is grey with yellow under the tail; the back and wings are dark. There are no recent records from Bali.

The fruit doves are mainly a Wallacean–Papuan genus; the Pink-necked is a montane endemic of Sumatra and Java, but the Black-naped has a wider range from Java and Nusa Tenggara to Sulawesi and the Philippines.

## LITTLE CUCKOO-DOVE
*Macropygia ruficeps* (30 cm)
**Kouran**                                                     Plate 6

As their name implies, these doves are cuckoo-like in shape, being slender and long-tailed. Not very easy to observe, they are most often seen in fast, direct flight just above or through the canopy. They occur in the forested areas of the mountains, but they fly quite long distances to feed and will occasionally descend to the adjacent plains. This species, the smallest of the three cuckoo-doves in Java, is mainly rufous in colour, darker above, and slightly metallic in the male. The call is a rapid 'croo-wuck, croo-wuck' repeated a few times. In the lowlands, it is replaced by a larger red–brown species, the BROWN CUCKOO-DOVE *Macropygia phasianella* (38 cm), which has a call consisting of a single rising and falling 'wow' at a rate of about 4 calls in 5 seconds.

## SPOTTED-NECKED DOVE
*Streptopelia chinensis* (30 cm)
**Tekukur**                                                Plate 6

The turtle-doves are also quite slender though they lack the very long tail of the cuckoo-doves. This familiar ground-feeding dove of open country and lower hills is often seen flying up from fallow fields and roadsides, even in towns. It is common at Jakarta Zoo. Watch for the black and white spotted half-collar and the broad white tips to the outer tail feathers. The call consists of three or four throaty cooing notes, from which it derives its local name. It is common from India to China and south to Java, but it is a popular cage-bird and has been introduced widely elsewhere.

## JAVAN TURTLE-DOVE
*Streptopelia bitorquata* (30 cm)
**Putar**                                                  Plate 7

This lovely dove has a more restricted range, in Java, Nusa Tenggara, and the Philippines. More of a woodland bird than the Spotted-necked Dove, favouring mangroves, and old coconut plantations, it does come out into open areas, and many can be seen feeding on the open grass around Soekarno-

Hatta Airport together with the Spotted-necked Dove. It differs from the latter in its much richer colour, with a narrow black half-collar bordered by white, and its very deep, slow, fruity 'cru-cruu' notes, so much an accompaniment to hot, drowsy afternoons on Pulau Dua.

## ZEBRA DOVE
*Geopelia striata* (21 cm)
**Perkutut**                                               Plate 7

This small, slender, long-tailed dove is better known to most people as a cage-bird, because excessive trapping has now restricted it to the wilder parts of Java and Bali. Its popularity stems from its musical call, a series of rapid trilled cooing notes. This dove prefers savanna and monsoonal woodlands, and the scrub growth at the back of sandy beaches. It is predominantly a ground bird, using low and partially hidden perches. The bird is named from the irregular black barring of the upperparts, neck, and flanks, but it is also known as 'Peaceful Dove'. An Australian species, it has extended north to the Philippines and southern Thailand.

# Parrots

EVERYONE is familiar with the great family of parrots which includes the cockatoos, lories, macaws, and parakeets, and many of the world's most colourful and friendly pet birds. However, parrots are poorly represented in Java, with only two species, a parakeet and a hanging parrot, although escaped Yellow-crested Cockatoos *Cacatua sulphurea* also occur wild in Jakarta Zoo. Parrots have heavy, hooked beaks used for tearing open fruit and flowers, often in a most wasteful fashion, and also used as a grip when climbing along branches. They nest in holes in the main trunks of trees, often using old nests of barbets or woodpeckers.

## RED-BREASTED PARAKEET
*Psittacula alexandri* (35 cm)
**Betet**                                                        Plate 7

Nearly half the length of this parakeet consists of tail. The
male is a beautifully coloured bird, as Plate 7 shows. A yellow
patch in the wing is prominent in flight. The female is slightly
duller. They live in active, noisy groups that dart and swerve
as they speed through the tree-tops, uttering raucous screams.
They are particularly noisy in the evening when flocks gather
to roost. They live in open, lightly wooded country, though
now they are common only in the more remote lowlands.
However, small parties still occasionally reach the countryside
near Jakarta (the author once saw a flock of 20 near Parung,
and there is a large colony at Jakarta Zoo). Outside Java, there
is a small population in South Kalimantan and on some West
Sumatran islands; its main range is from India to China.

## JAVAN HANGING PARROT
*Loriculus pusillus* (14 cm)
**Serindit**                                                     Plate 7

This tiny bundle of whirring green, quite unlike the parakeets,
is usually seen as a flash past the tree-tops; it is a deep green bird
with an almost headless appearance, stumpy wings and short
tail, uttering a thin, high 'zri-ie'. Active little birds, they
clamber about the branches with great dexterity and always
rest or sleep hanging upside-down. It is more a matter of luck
than of skill to obtain a good view, but the species does not
appear to be very common. It is generally considered an endemic in Java and Bali, although very similar to another form
occurring in the oriental region, while a related species is
found in Sumatra, Kalimantan, and Malaysia.

# Cuckoos

THE true cuckoos are well known for their selfish, lazy life-style of laying their eggs in the nests of other birds and leaving the rearing of their young to the foster parents ('brood parasites'). Being primarily arboreal, many cuckoos are difficult to observe, but all have distinctive songs which often reveal their presence. Many cuckoos habitually sing at night. They have rather long, generally barred tails, and a slightly down-curved bill. The family includes two quite distinct, non-parasitic groups, the malkohas and coucals.

## PLAINTIVE CUCKOO
*Cacomantis merulinus* (22 cm)
**Wiwik Kelabu**                                      Plate 7

The Plaintive Cuckoo is a familiar bird, thanks to its voice. Found in lightly wooded open country, it favours low perches and even telephone wires and fences, so is not too difficult to observe. It can be heard in town gardens, even occasionally in central Jakarta. It is quite small for a cuckoo. The adult plumage in Plate 7 would appear to be distinctive, but the Indonesian Cuckoo is very similar (see below), and the best identification is from its high-pitched whistled song. There are two quite separate songs which can be interchanged at will. The less frequent song consists of the phrase 'tay-ta-wi' repeated about four times up the scale, but the cadence song is diagnostic: it begins with four slow notes, breaking into eight rapid notes down the scale. The usual foster parents are believed to be the small ioras and warblers.

Brief notes are given on the other cuckoos which the keen observer should find in Java:

## INDONESIAN CUCKOO
*Cacomantis sepulcralis* (24 cm)

Known also as Brush Cuckoo, this bird is darker than the Plaintive Cuckoo, with the head usually as dark as the back. The rising song is almost the same, but the cadence call differs in consisting of a long series of at least ten plain whistles, 'heet, heet, heet, . . .', at the same speed, falling very slightly in pitch. It lives in forest margins and well-wooded country, but is also sometimes found in rubber estates and villages, even on the outskirts of Jakarta.

## BANDED BAY CUCKOO
*Cacomantis sonneratii* (23 cm)

This species is brown above and whitish below, but with fine black bars over the entire body (but it should be noted that the young birds of most cuckoos are also barred). The diagnostic call consists of four thin, shrill whistles falling slightly but evenly in pitch. It lives in well-wooded country and forests, and is not very common. It does not occur in Bali.

## ORIENTAL CUCKOO
*Cuculus saturatus* (33 cm)

A large cuckoo of mountain forests, this is commonly heard but rarely seen. The call is three fast mellow 'hoops', the first pitched higher and not heard at a distance, so that often only a double 'hoop-hoop' is heard.

## DRONGO CUCKOO
*Surniculus lugubris* (23 cm)

This is a black drongo-like bird, even to the extent of having a very slightly forked tail, but unlike a drongo, the under tail coverts are strongly barred with white. The song is a human whistle of five to seven even notes rising up the scale. Not very common, it lives in well-wooded country and forests; there are no recent records from Bali.

# KOEL
*Eudynamys scolopacea* (43 cm)

A large, heavy, but secretive cuckoo, black in the male and brown with pale markings in the female. The call consists of a loud 'ko-el' rising in pitch and crescendo; it also makes excited bubbling calls. It is parasitic on crows, and its presence is now confined to the more remote wooded areas of Java.

# CHESTNUT-BREASTED MALKOHA
*Phaenicophaeus curvirostris* (47 cm)
**Kadalan**                                               Plate 8

Although malkohas belong to the cuckoo family, they appear to have little in common. They rear their own young and are mostly silent. They are large, long-tailed birds (the length includes 25 cm of tail) that creep around dense foliage almost like squirrels, occasionally making short, rather ungainly flights to the next bush on broad, rounded wings. As there are only two species on Java (compared to six on Sumatra), they are easy to identify. The Chestnut-breasted Malkoha is distinguished by its thick green bill and the lack of white in the tail. This malkoha occurs in densely wooded localities in the lowlands and lower hills.

The RED-BILLED MALKOHA *Phaenicophaeus javanicus* differs mainly in having a red bill and white bars on the underside of the tail. The breast is a softer, paler chestnut. It is not very common, but ranges up to a height of at least 2 000 m on the Ijen Plateau. It does not occur in Bali.

# LESSER COUCAL
*Centropus bengalensis* (40 cm)
**Bubut Alang-alang**                                       Plate 8

The coucals or crow-pheasants are heavy, rather untidy, thick-billed black birds with moderately long graduated tails and chestnut wings. In non-breeding and immature plumage, the Lesser Coucal is dark brown rather than black, much spotted and streaked. It hops around the scrub in open country, occasionally making short, ungainly flights. Its frequently heard call consists of three or four hollow notes breaking into an unmusical staccato call: 'boob, boob, boob, kok-ok-oo, kok-ok-oo, kok-ok-oo'. Another call consists of rapid notes increasing in tempo and falling in pitch.

In more densely wooded country and forests, the GREATER COUCAL *Centropus sinensis* (52 cm) replaces the Lesser; it is a cleaner-looking bird, but quite secretive. The call is a deep, hollow 'boob' repeated four or five times. The full song, which is not often heard, consists of a prolonged series of such notes lasting up to 15 seconds, beginning rapidly but tailing off into many slow 'hoops'.

The SUNDA COUCAL *Centropus nigrorufus* (45 cm) is a rare endemic on Java. There is one old record from Sumatra, but as this is held in doubt, perhaps JAVAN COUCAL is a more appropriate name. Intermediate in size between the other coucals, it differs in that its entire body is black, including the back, which is chestnut in the other species. The chestnut flight feathers are also edged a very dark brown, and there is a purple gloss on the head and neck. The call is said to be very similar to that of the Greater Coucal, but further work is required to confirm this. The main habitat appears to be tidal forests, which are now rare in Java, and the only recent records are from the mangroves around Cilacap, where it might still be quite common. It should be looked for elsewhere, however, especially in Ujung Kulon.

# Owls

IT is a pity that owls are treated with such superstitious fear by many people, for their faces and mannerisms are a delight to observe. Their flat facial discs sharpen their hearing, their large round eyes are an aid to nocturnal vision, and they have the ability to turn their heads through an angle of 180°. They are mainly nocturnal, and daytime activity will often result in frantic mobbing by a host of small birds, even though owls may not be harbouring evil intentions against them at that time.

## BARN OWL
*Tyto alba* (34 cm)
**Serak**

Plate 8

Although believed to be quite widespread in open country, this owl is little known in Java and is rarely seen. It has a nearly world-wide distribution, and acquires its name from its habit of roosting and nesting in old buildings. In Java, however, it is believed to use mostly holes in large trees. Whitish in colour with a grey and buff tinge and markings, it has a flat, heart-shaped facial disc. It hunts on completely silent wings, and its appearance low over the ground at dusk is indeed ghostly, while its call is an eerie, long shrilling sound. The call can be heard, though rarely, over open ricefields and villages, and sometimes—in the middle of the night—even over Bogor or the outskirts of Jakarta. Watch for its ghostly figure at floodlit events in rural areas.

There are two large forest owls in Java, but they are now scarce. The BARRED EAGLE-OWL *Bubo sumatranus* (45 cm) is blackish-brown barred with buff, and lives in deep forest, in both the lowlands and the hills. The BUFFY FISH-OWL *Ketupa ketupu* (45 cm) is rufous, with black streaks (not bars) and unfeathered legs; it is nearly always found near water. Both have prominent ear tufts, though these are not always held erect.

## COLLARED SCOPS OWL
*Otus bakkamoena* (22 cm)
**Celepuk** Plate 8

The commonest and most widespread in Java, this little owl can often be heard calling from tree-lined avenues in large towns and in gardens, villages, and the wooded areas of the hills. The call is a single 'wok', slightly rising, uttered at irregular intervals for long periods. This owl is brownish in colour, lightly streaked and mottled, with a bright half-collar on the back of the neck, and small ear tufts.

In wooded areas, especially in southern Java, listen for the JAVAN BARRED OWLET *Glaucidium castanopterum* (23 cm), a species endemic to Java and Bali. Typically it calls for about 15 minutes, 10 minutes after Maghrib (evening prayers) and again at first light, though on occasions it will call through much of the night. It is a strange and eerie trilled note, repeated in series and gathering speed and volume until it ceases abruptly, to begin again a minute or two later. To the uninitiated, this could easily be mistaken for a barbet call, but the latter is more mellow in tone and lacks the owl's haunting quality.

In the mountains of West Java, the JAVAN SCOPS OWL *Otus angelinae* (20 cm) must be one of Indonesia's least-known endemics, recorded only from the forests of one or two mountains, such as Gunung Gede and Gunung Pangrango, at 1 400–2 000 m. Most owls have distinctive calls, but this one appears to be habitually silent, which explains why it is so poorly known. However, two young birds on Gunung Pangrango were observed giving the typical hunger or contact calls of young owls in February 1985. The bird's conspicuous features are the golden iris and the white eyebrows which extend into short ear tufts. It is possibly not uncommon within its restricted range, but lucky indeed is the bird-watcher who can ever find one except through the use of nets.

# Nightjars

NIGHTJARS are nocturnal birds that have long, slender, pointed wings, long tails, and rather hawk-like flight. Their plumage consists of beautifully mottled shades of black, browns, and grey, almost impossible to describe but an excellent camouflage when the bird rests during the day. Most nightjars have some white on the wing and throat. Their bills appear small but open to reveal a huge gape by which they catch insects in flight. They often rest on bare ground and even on roads, where their reddish eyes reflect car headlights at night.

## SAVANNA NIGHTJAR
*Caprimulgus affinis* (23 cm)
**Cabak Maling**                                          Plate 9

This is the nightjar of open ground and scrub, as often occurs behind beaches, for example, but it has also adapted well to a completely urban environment, and its slightly querying 'schwick' calls are a nightly accompaniment to the traffic noises of downtown Jakarta. It is still not known where the urban birds roost by day, but patches of waste ground and the flat tops of buildings have both been suggested. However, numbers have declined in Jakarta in recent years, probably as a result of waste plots being developed. When seen at close quarters, the male (only) has diagnostic white outer tail feathers. This species ranges from India to southern China and the Philippines, Sulawesi, and Nusa Tenggara. It is not found in Malaysia, but it has recently been reported that a small colony has become established in Singapore. If that colony grows, it will become an attractive addition to that urban island's avifauna.

In more wooded country, the LARGE-TAILED NIGHT-JAR *Caprimulgus macrurus* (30 cm) may be heard, identified by its monotonous loud, slightly irregular but persistent 'chonk, chonk' calls at dusk and sometimes through the night. This species is most common in the wilder parts of Java, but both can be heard at Muara Angke, close to Jakarta. This nightjar does not usually roost on the ground, preferring to lie lengthwise along branches.

# Swifts

SWIFTS are aerial insect-eaters identified by their long, slender, often scythe-like wings and predominantly black or dark brown plumage. Some species are believed to be entirely aerial, even sleeping on the wing and alighting only to breed, and although this may not apply to any of the resident swifts of Java, some will spend the entire day on the wing, travelling great distances from their roosting sites. Some groups, particularly the needletail swifts and swiftlets, are notoriously difficult to identify, even for the experts. The taxonomy of the swiftlets is still being unravelled, and there could be one or two species endemic to Java, depending upon their final taxonomic classification, but such problems have no place in this book. Some swiftlets use echo-location for navigation in caves and are the source of bird's nest soup, the nest being constructed from the bird's own saliva. One should be careful not to confuse swifts with swallows, which do not have such long, narrow wings.

HOUSE SWIFT
*Apus affinis* (15 cm)
**Kapinis Rumah** Plate 9

Government employees in Bandung know this bird only too well, for it breeds in hundreds not merely on the Gedung Sate office building but also inside it, where constant chittering calls accompany the clerks along the corridors. Swifts of this genus are not as small and slender as the swiftlets; they have slightly forked tails, though this is not always obvious. The rump and throat are white. This species occurs widely from Africa to Java, Sulawesi, and the Philippines, but its distribution in Java is rather patchy. It appears to be only a casual visitor to Jakarta.

## PALM SWIFT
*Cypsiurus balasiensis* (13 cm)
**Burung Kendali**                                    Plate 9

This distinctly slender swift has long, narrow, curved wings and a long, narrow, forked tail, and is dark brown without a white rump. It breeds in palm trees in cultivated areas, especially in the borassus palms of lowland ricefields, and occasionally reaches the outskirts of towns. It occurs from India and South China south to Java and the Philippines, and has recently been found also in southern Sulawesi.

## LINCHI (WHITE-BELLIED) SWIFTLET
*Collocalia linchi* (10 cm)
**Walet Sapi**                                        Plate 9

The swiftlets are a very difficult group to identify, as they are small and lack distinctive plumage. Some are nearly as slender as the Palm Swift. However, this tiny bird, the smallest of the family, is readily recognized by its wholly white belly and glossy green upper-parts, though they usually appear blackish. Workers in tall office buildings in Jakarta will often see them riding the air currents, hawking for insects outside their windows. They breed in culverts, bridges, and buildings throughout Java, Bali, and Lombok, and this or a very similar form occurs widely in South-East Asia.

Two species produce edible nests in Java: the EDIBLE-NEST SWIFTLET *Collocalia fuciphaga*, which produces the valuable white nests made almost entirely of saliva, and the BLACK-NEST SWIFTLET *Collocalia maxima*, whose nest consists of saliva mixed with its own feathers. They breed typically in caves, rarely in houses, and use rattling calls as a means of echo-location in the dark, in a sort of avian radar. Both birds are dull black. Two main breeding centres of the Edible-nest Swiftlet in Java are Gresik and Rembang.

# Trogons

THE trogons, shy denizens of the middle storey of dense forest, are some of the most colourful birds in Java. They sit quietly, singly or in loose pairs, and are easily overlooked, unless the low 'prrt' calls or whirring of wings, as they sally forth after an insect, attract attention. They never make long flights or leave the forest. There are two species in Java (compared with seven in Sumatra), but neither occurs in Bali.

BLUE-TAILED TROGON
*Harpactes reinwardtii* (33 cm)
**Burung Luntur**                                        Plate 11

The Blue-tailed Trogon is a montane endemic of Java and Sumatra, and can sometimes be seen at Cibodas. The male is unmistakable with its bright colours and red bill, but often, all that an observer sees is a dark-backed bird with white tips and borders on the outer tail feathers as it flies off to a new perch just out of sight.

In lowland forest, and therefore now scarce, one may find the ORANGE-BREASTED TROGON *Harpactes oreskios* (30 cm), which has a greenish head and orange belly (yellow in the female), and blue bill. It does not normally reach the altitude of Cibodas. The call is a rapid series of loud 'chaw' notes.

# Kingfishers

EVERYONE knows the kingfishers with their colourful plumage, large, dagger-like bills, and habit of plunging into water to catch fish. However, there are some species, in both forest and open country, which are not associated with water, feeding instead mainly on lizards and beetles. Kingfishers are shy birds, most often seen in swift and direct flight away from the observer, though the larger species can be very raucous.

## SMALL BLUE KINGFISHER
*Alcedo coerulescens* (14 cm)
**Burung Udang**                                               Plate 10

This small kingfisher has soft greenish-blue upper-parts and white under-parts with a distinctive light blue breast band. There is no red in the plumage. A bird of the coasts, it is found equally on sandy beaches, mangrove edges, and in swamps and ricefields of the coastal plains. It perches on bushes, earth banks, and man-made structures, but never high in trees. It feeds on fish, hovering briefly before diving. The call note is a soft 'tit' or 'tit-ti-tit' given in flight. It is endemic to Java eastwards to Sumbawa, but has recently been discovered in Lampung and South Sumatra, where it could be extending its range.

## DEEP BLUE KINGFISHER
*Alcedo meninting* (15 cm)                                     Plate 10

Plate 10 shows that this is a much more deeply coloured bird than the Small Blue Kingfisher, with a vivid blue back and brick-red under-parts. Primarily a bird frequenting streams in thick woodland in Sumatra, in Java it also frequents more open country such as the rivers flowing through the Jakarta Zoo or beside ricefields. Generally, it is seen as a ball of deep blue darting along a river with a quiet but high, sharp squeak. There appear to be no recent records from Bali.

## JAVAN KINGFISHER
*Halcyon cyaniventris* (26 cm)
**Cekakak**                                    Plate 10

As implied by the name, this kingfisher is endemic to Java and Bali. The purple plumage with vivid blue flight feathers and tail is distinctive, as is the huge red bill. A white patch in the wings is seen in flight. Although a bird inhabiting more open country, it usually avoids the larger swamps and only incidentally occurs near rivers. It is a shy bird, quite difficult to see, but the call is a very loud screaming chatter, identical to that of its near cousin, the WHITE-BREASTED KING-FISHER *Halcyon smyrnensis*, which occurs widely in Sumatra and mainland Asia, and which has been reported from West Java.

On the coast, look out for the huge STORK-BILLED KINGFISHER *Pelargopsis capensis* (37 cm), with its massive red bill. The head and body are brown, but the upper-parts and wings are blue, with the usual brilliant blue down the lower back. It can be extremely noisy but, sadly, it is now rare in Java and may be found only on the more remote mangrove coasts.

## COLLARED KINGFISHER
*Halcyon chloris* (24 cm)
**Cekakak**                                    Plate 10

Even this noisy kingfisher, so common on coasts from India to Australia, has become somewhat local in Java, where it is probably a victim of airguns and catapults. Its habit of sitting on exposed perches makes it an easy target for such destructive sport. Its favoured habitat is coasts, where strips of casuarinas behind the beach may ring with its screaming and rather irritating 'kek-kek, kek-kek-kek' calls. It often perches on isolated rocks or stumps on the tidal flats. It also ranges widely inland, in golf-courses and gardens, reaching even Cibodas and the Ijen Plateau. The blue–green crown and back, blue wings, and white collar and under-parts are distinctive.

During the Australian winter, from April to September, one should be careful not to confuse it with the very similar migrant SACRED KINGFISHER *Halcyon sancta* (20 cm). This bird is not only smaller but also duller, with a darkish tinge above and a buff tinge below. Found in similar habitats, it seems to be a more familiar bird, less wild and raucous, and regularly using lower perches in *ladang* (arable plots in wooded areas) and gardens.

# Bee-eaters

BEE-EATERS are graceful, slender and quite colourful birds with rather long, slightly curved bills. They feed on airborne insects, though some species prey largely on bees and have learnt how to deal with the stings. Like swifts and swallows, their flight can be quite acrobatic.

## CHESTNUT-HEADED BEE-EATER
*Merops leschenaulti* (22 cm)
**Kirik-Kirik**                                      Plate 11

This small bee-eater is usually found in small parties on woodland edges, including teak, eucalyptus, and cocoa plantations. The beautiful plumage includes a pale blue rump which is distinctive in flight. The call is a soft 'kwlp, kwlp'. It feeds mostly by sallying forth after insects from an exposed perch, in contrast to the larger BROWN-BREASTED BEE-EATER *Merops philippinus* (30 cm), which feeds mainly by hawking in flight over open country. This latter bird is mainly green, with yellow and rufous on the throat when seen at close range, but it also has the blue lower back. It is a common migrant from the north, with louder call notes which are often heard from parties migrating overhead, especially from March to May and from September to October.

# Hornbills

EVEN the least interested visitor to Indonesian forests will enquire about the identity of these large, obtrusive birds which are, unfortunately, no longer a very characteristic part of the Javanese scene. They have, for example, not been seen at Cibodas for many years, and there are very few in Bali. The name is derived from the huge bill which carries a horny protuberance or casque on the upper surface. Hornbills are fruit-eaters and often fly quite long distances in their search for forest trees in fruit, especially the wild fig. There they will gather with barbets and pigeons, such gatherings always being a memorable occasion for the bird-watcher. Their wings are extraordinarily noisy in flight, with a 'whooshing' sound that has been likened to that of a steam engine. They breed in holes in the main trunk of trees, and the female becomes a prisoner while incubating by partially sealing up the hole with mud for protection against predators such as monkeys. Probably the best place to see hornbills in Java is at Meru-Betiri.

# WREATHED HORNBILL
*Rhyticeros undulatus* (100 cm)
**Julang** Plate 11

The Wreathed Hornbill is black with a white neck and tail, though these are often stained yellowish. The small crest is chestnut, and the casque is comparatively small. The male has a yellow pouch on the throat, crossed by a narrow black band. It has a double barking call, but it is not often heard. It lives in the forested areas of the hills and the more densely wooded parts of the southern lowlands. The author once saw a party of three flying over the populous lowlands near Jember, from one wooded hill to another.

There are two other hornbills in Java. The large RHINO-CEROS HORNBILL *Buceros rhinoceros* ('Rangkong' in Indonesian) measures 120 cm for the male, but only 90 cm for the female. It has a large casque, a black band in the white tail, and a loud trumpeting call. It is now rare.

The smaller PIED HORNBILL *Anthracoceros coronatus* ('Kangkareng' in Indonesian) is 75 cm; it is black with a white belly, a white wing bar, white outer tail feathers, and large white casque. Its shrill piping calls are mostly heard from small parties in lowland woods, especially near water. The Pied Hornbill is still quite common in remote areas such as Ujung Kulon, Pangandaran, and Baluran.

# Barbets

BARBETS are familiar to everyone who has walked in forest or woodland, as their incessant calls are a constant accompaniment. However, they spend long hours perched in the upper canopy, where their green plumage blends with the foliage, and they are difficult to spot. Fruit-eaters, they are rather stout birds with heavy bills, and sometimes many barbets will attend fruiting fig trees in the forest. Their flight is direct and swift, on whirring wings, but they rarely fly very far. Most species have very brightly coloured heads which should be studied carefully when one is lucky enough to get a view. All have characteristic calls, but some have two or more quite distinct types of call. Once learnt, most of the calls are diagnostic.

## COPPERSMITH BARBET
*Megalaima haemacephala* (15 cm)
**Ungkut Ungkut**                                           Plate 12

The Coppersmith is the smallest of the Indonesian barbets. The body is green, but the belly is yellowish with green streaks. The forehead, eye patches, throat, and breast are crimson, as are the legs. (Sumatran birds differ in having a yellow throat and breast with a crimson bar across the upper breast.) The call, a regular 'tonk, tonk, tonk' from which it gets its name, is kept up for long periods at the rate of slightly more than one 'tonk' per second. The Coppersmith is a bird of open woodland which can include villages and groves of trees in open cultivation. Sometimes it is heard in the Bogor Botanical Gardens. Widespread in the Orient, its range extends from Pakistan to South-west China, south to the Philippines, Java, and Bali, but excluding Borneo and Sulawesi. It is found mainly in the lowlands and lower hills, though it is present on the Ijen Plateau.

There are five other barbets in Java, three of which are endemic. The Coppersmith is fully illustrated in Plate 12 but only the head patterns of the others are shown. The calls and habitats are described below.

## BLUE-EARED BARBET
*Megalaima australis* (18 cm)
**Tengeret**                                                Plate 12

The call is a constant disyllabic note, 'tk-trrk', repeated about twice per second, sometimes for much of the day. A second call is monosyllabic, similar to a referee's whistle, uttered at the same speed or faster than the first, but for shorter periods. Found mainly in the forests of the lowlands, this barbet also extends into the hills. It has a green body, bluish tail, and a yellow and blue head pattern. In Sumatra and Kalimantan, they have red patches on the cheeks.

## BLUE-CROWNED BARBET
*Megalaima armillaris* (19 cm)
**Tohtor**                                                  Plate 12

The call is represented as 'trrk-trk-trrrk', or three to five short trilled notes followed by a long rolling note. Found in mountain forests, this species is endemic to Java and Bali.

## BLACK-BANDED BARBET
*Megalaima javensis* (25 cm)
**Tulung Tum Puk**                                          Plate 12

The typical call is 'toolung-tupuk' or 'tutuk-tuk-tuk' repeated at intervals. This barbet is found in forest and woodland mainly in the lowlands, although it also occurs on the Ijen Plateau. It is endemic to Java, but its presence on Bali requires confirmation.

## LINEATED BARBET
*Megalaima lineata* (28 cm)
**Bultok**                                                  Plate 12

An emphatic 'bul-tuk', the second note sharper and a bit higher, is repeated in phrases or at irregular intervals. Occasionally, the bird gives a series of more musical trilled notes, beginning with several long trills followed by a series of four-syllable notes. A bird of lowland, lightly wooded terrain, including cultivation, it is probably most common in the monsoon forests of the east, where it extends up to about 1 500 m. It is found from Peninsular Malaysia to India and China, with an outpost in Java and Bali.

## BROWN-THROATED BARBET
*Megalaima corvina* (26 cm)
**Bututut**                                               Plate 12

This is a montane endemic confined to West and Central Java. Its calls are more varied, usually a series of 'tuk, tuk, tuk' notes with slightly irregular deliverance, sometimes breaking into a rattle. It has also a trilled note similar to that of the Lineated Barbet.

# Woodpeckers

THERE are 15 species of woodpecker in Java, but as most of these favour lowland forests, they have become rather local in distribution. They range widely in size, from 9 to 50 cm, and also in general habits and colour, but all are characterized by the ability to climb vertically up the trunks and branches of trees, pecking vigorously at the bark in their search for insects. For this purpose, they have long, sharp bills and specialized feet and tails, the latter acting as a support. Their dipping flight, with rounded wings and rather stumpy rear quarters, is characteristic of the family. Their calls, which are often sharp or harsh, particularly in the case of the larger species, are not usually a great help in identification. Some species also 'drum' on wood very rapidly with their bills, as another form of vocalization. For nesting, they bore a hole into the rotten centre of old or dying trees.

# CRIMSON-WINGED WOODPECKER
*Picus puniceus* (25 cm)
**Celadi, Pelatuk**                                          Plate 13

This is quite a common forest bird of both lowlands and mountains, but it is absent on Bali. Although colourful, as shown in Plate 13, it is not very easily seen, except as a silhouette high up the trunk, or settled for the night at the very top of a tree. When you get a good view, record the colours carefully, particularly in the head and neck region, and the extent of barring on the body, in order to distinguish it from a number of related species. The call, a rather shrill 'sheok', is uttered once or twice or as a run of several notes descending slightly in pitch.

A more striking forest woodpecker is the ORANGE-BACKED *Reinwardtipicus validus* (30 cm). It has an elongated look, enhanced by an orange–yellow stripe (whitish in the female) along the length of its back. Chestnut bars in the black wings and, in the male, a red crest and under-parts combine to make it a distinctive bird. It can be observed readily as it generally keeps to the lower branches, often in noisily chittering family parties. It occurs in the lowlands and hills of Java, but not Bali.

Formerly, the common large woodpecker of the open lowlands was the COMMON GOLDEN-BACKED WOODPECKER *Dinopium javanense* (30 cm), but this beautiful bird has become quite scarce except in the more remote districts. The author once found it very common on an estate of old coconut trees on the south coast. The mantle and wings are golden yellow, the lower back is red, the head region has broad black and white stripes, and the under-parts are white marked with heavy black scales. The male has a red crown and pointed crest. Usually, all one sees of this bird is a flash of gold as it flies away from the trunk of a tall tree, often a coconut palm, with a loud metallic trilling call and heavy dipping flight. The beginner should be careful not to confuse this bird with another golden yellow bird, the Black-naped Oriole.

## FULVOUS-BREASTED WOODPECKER
*Dendrocopus macei* (18 cm)                                    Plate 13

This small woodpecker is quite common in open and lightly wooded country, favouring, for example, scattered trees in golf-courses, but it is widely distributed and can be seen even in suburban gardens and the forest margins of the hills. In flight it looks mainly brown, but a closer view will reveal barred black and white upper-parts and mainly buff, faintly streaked under-parts. The cheeks are a clean white, bordered below by a heavy moustachial stripe. The crown is red in the male, black with a red forehead in the female, and both sexes have light red under the tail. The call is a soft 'plic' or metallic trill that often reveals the bird's presence.

# Pittas

## BANDED PITTA
*Pitta guajana* (23 cm)
**Burung Paok**                                                Plate 13

It is very difficult to see this bird in the wild, but it is quite common in damp patches of forest in the lowlands and lower hills. This is characteristic of pittas: although very colourful, they are skulking birds of deep shady cover, and are almost exclusively terrestrial, seen singly or in pairs. The art of finding pittas, and other secretive forest birds, is to learn their calls, and the bird-watcher with an enquiring mind must have the time and patience to follow up any calls that he does not know. This pitta has a short, sharp 'pow' or 'piau' note and a low, slightly falling purr, 'crrrrr'. With luck and stealth, one can sometimes get short glimpses as it walks through the undergrowth or jumps briefly, with a hopping gait, on to a fallen log. To have one's patience rewarded with a good view of a pitta is one of the most satisfying experiences a bird-watcher can have.

# Larks

SINGING BUSHLARK
*Mirafra javanica* (15 cm)
**Branjangan**                                              Plate 13

A rather small, dull-coloured bird of open country, the bushlark is readily overlooked. It is, however, common in the plains, including riceland, dryland cultivation, and grassland. Of rather stocky, short-tailed appearance, it has the brown-streaked plumage common to most larks and pipits, but if encountered at short range, look for rufous in the wing. The bushlark is most readily recognized by its songflight, a slow fluttering flight on rounded wings at a height of about 10 m, during which it sings continuously a series of slow, sweet trills and single notes.

# Swallows

SWALLOWS are familiar to most people, noticed for their elegant shape, graceful flight, friendly twittering calls, and frequent proximity to man. In contrast to swifts, they perch frequently, and lack that family's scythe-like, long, pointed wings.

# BARN SWALLOW
*Hirundo rustica* (15 cm, or 20 cm with tail streamers)
**Layang-Layang**                                          Plate 13

This species breeds nearly world-wide in the northern hemi-
sphere, migrating south in winter, and although only a migrant to Java, it is abundant everywhere between the months
of August and April, with some lingering throughout the year.
In its breeding grounds, its return is eagerly awaited as a
harbinger of spring after the long winter months.

It is a slender bird with pointed wings, and the outer tail
feathers are elongated as pointed streamers. The tail is fanned
on perching, or veering in flight, to show white streaks along
the tips of the shorter feathers. This swallow is glossy blue
above, and whitish below, with a chestnut throat demarcated
from the breast by a narrow dark band. There is also a small
chestnut patch on the forehead.

Migrating and wintering swallows commonly perch in
dense, twittering rows along telephone or electric lines in
towns or along roads, and the birds of an entire district will
concentrate at dusk in certain towns. The author has seen such
roosts, sometimes numbering thousands of birds, in many
towns in Sumatra and Kalimantan, and is told that there is one
in Yogyakarta. Swallows feed by catching insects in swift
flight with a high degree of agility, swooping and swerving
low over the ground.

Swallows should not be confused with swifts, which have
longer, narrower, and more crescentic wings and generally
dull, dark plumage. More particularly, in the autumn especially, a great many Barn Swallows are immature, with
dirtier plumage and without the elongated tail feathers, and
these swallows can be confused easily with the resident
PACIFIC SWALLOW *Hirundo tahitica*. In this latter species,
the chestnut throat is more extensive, lacking the dark
bordering breast band, the under-parts are greyish, and the

outer tail feathers are only slightly elongated. The Pacific Swallow breeds mostly on buildings, constructing mud nests in the corners between wall and ceiling, usually on the inside of house verandahs. It is common everywhere, with a pair or two in most villages.

A larger bird occasionally encountered in open country, or in the river gorge that flows through Bogor, is the RED-RUMPED SWALLOW *Hirundo striolata*. This heavier species, identified at once by its pale chestnut rump, has a slower flight. Its under-parts are white streaked with black.

# Cuckoo-Shrikes

CUCKOO-SHRIKES are related neither to cuckoos nor shrikes but they have some superficial resemblances to both groups, and were so named when early ornithologists in tropical regions were at a loss for names when faced with groups of birds totally unfamiliar to them. The family includes a rather motley group of mainly arboreal birds, of which only the more common and distinctive are included here.

### WHITE-WINGED TRILLER
*Lalage sueuri* (17 cm)
**Kapasan**                                          Plate 14

The trillers are medium-sized birds with rather broad but pointed wings and moderately long but rounded tails. They are strongly patterned, black and white in the male, brown and dirtier white in the female. The black of the male is glossy, but the lower back is grey, and the under-parts, eyebrow, outer tail feathers, and broad patches in the wings are white.

The triller is seen singly, in pairs, or small family parties, in a wide range of habitats from mangroves, secondary growth, and open country to forest margins and estates. While these birds may perch on the exposed branches of trees and on telegraph poles, and occasionally even the ground, they are not always easy to see in the dense foliage of a tree. The White-winged Triller is the author of the rather loud trill—a rapid musical, but slightly metallic, whistle or chatter on one pitch—that is one of the characteristic sounds of lowland East Java during the dry season. This is an Australian species that extends to East Java and Sulawesi.

The PIED TRILLER *Lalage nigra* (16 cm) has a Malaysian distribution, south to Central and West Java. It is a smaller but otherwise almost identical bird, except that the female is darker, greyish-brown. It is a more familiar bird of gardens and moist woodlands and has a softer and sweeter song. The two species overlap in East Java, Sulawesi, and perhaps also in Bali, where the Pied Triller seems to prefer the moister localities such as wooded swamp margins.

## BLACK-WINGED FLYCATCHER-SHRIKE
*Hemipus hirundinaceus* (17 cm)                    Plate 14

Found in forests and forest margins from the lowlands to an altitude of about 1 500 m, this small bird lives in parties in the upper canopy and would not often be noticed except when its twittering notes attract attention. It is not difficult to identify, being all black above except for a white rump, and all white below. In the female, brown replaces the black.

The cuckoo-shrikes proper are usually larger birds of varying shades of grey. The BLACK-FACED CUCKOO-SHRIKE *Coracina larvata* (25 cm) is grey, with a black face and tail, and white under the tail. It is most often encountered in pairs on the margins of montane forests, though it will also join mixed parties of birds in the forest.

# Minivets

ALTHOUGH minivets belong to the cuckoo-shrike family, there is little obvious similarity. Minivets are small, brightly coloured, long-tailed birds that live mainly in the canopy in cheerful, twittering parties that sometimes number over 30 birds. Commonly, individual birds will fly from tree to tree in progression, but the loose party always maintains contact through their soft, high-pitched calls, which are difficult to describe but, once known, easy to recognize. There are only three species in Java.

## SUNDA MINIVET
*Pericrocotus miniatus* (19 cm)
**Burung Sepah Gunung**                                    Plate 14

Quite common, this minivet is endemic to the montane forests of Java and Sumatra (though not Bali) and can be seen often at Cibodas, for example, in rather large parties in the canopy or in overhead flight, their calls always attracting attention. The tail is noticeably long, and the male is a lovely bird with its glossy black and deep red plumage. The female is also red, though paler than the male.

The male SCARLET MINIVET *Pericrocotus flammeus* is very similar, but it is mainly a bird of lowland forest. However, it does occur in the hills even to the height of Cibodas, so some care is required. Usually the presence of the female will immediately establish identity, for she has the flame-red parts of the male replaced with bright yellow.

## SMALL MINIVET
*Pericrocotus cinnamomeus* (16 cm)
**Burung Sepah Kecil**                                      Plate 14

The Small Minivet frequents open cultivation and lightly wooded areas mainly in the lowlands, and sometimes comes

9. (a) House Swift *Apus affinis*. (b) Palm Swift *Cypsiurus balasiensis*. (c) Linchi Swiftlet *Collocalia linchi*. (d) Savanna Nightjar *Caprimulgus affinis*.

10. (a) Javan Kingfisher *Halcyon cyaniventris*. (b) Collared Kingfisher *Halcyon chloris*. (c) Deep Blue Kingfisher *Alcedo meninting*. (d) Small Blue Kingfisher *Alcedo coerulescens*.

11. (a) Wreathed Hornbill *Rhyticeros undulatus*. (b) Blue-tailed Trogon *Harpactes reinwardtii*. (c) Chestnut-headed Bee-eater *Merops leschenaulti*.

12. (a) Brown-throated Barbet *Megalaima corvina*. (b) Coppersmith Barbet *Megalaima haemacephala*. (c) Blue-crowned Barbet *Megalaima armillaris*. (d) Black-banded Barbet *Megalaima javensis*. (e) Blue-eared Barbet *Megalaima australis*. (f) Lineated Barbet *Megalaima lineata*.

13. (a) Crimson-winged Woodpecker *Picus puniceus*. (b) Fulvous-breasted Woodpecker *Dendrocopus macei*. (c) Barn Swallow *Hirundo rustica*. (d) Banded Pitta *Pitta guajana*. (e) Singing Bushlark *Mirafra javanica*.

14. (a) White-winged Triller *Lalage sueuri*. (b) Black-winged Flycatcher-Shrike *Hemipus hirundinaceus*. (c) Common Iora *Aegithina tiphia*. (d) Sunda Minivet *Pericrocotus miniatus*. (e) Small Minivet *Pericrocotus cinnamomeus*.

15. (a) Sooty-headed Bulbul *Pycnonotus aurigaster*. (b) Yellow-vented Bulbul
*Pycnonotus goiavier*. (c) Orange-spotted Bulbul *Pycnonotus bimaculatus*.
(d) Black-headed Bulbul *Pycnonotus atriceps*.

16. (a) Black Drongo *Dicrurus macrocercus*. (b) Lesser Racket-tailed Drongo *Dicrurus remifer*. (c) Large-billed Crow *Corvus macrorhynchos*. (d) Black-naped Oriole *Oriolus chinensis*.

into villages and even wooded suburbs on the outskirts of Jakarta. The male has rather softer colours than the other minivets, being grey and orange with black and orange wings and tail, while the female is greyer, with yellow replacing the orange on the wings and rump. In its unobtrusive way, it is one of the most attractive birds of lightly wooded low country. It occurs in India and mainland South-East Asia, though it is replaced by a related red species in Sumatra, Kalimantan, and Malaysia.

# Ioras and Leafbirds

## COMMON IORA
*Aegithina tiphia* (15 cm)
**Pito** or **Cipoh**                                    Plate 14

Another favourite among the birds of lightly wooded country is the iora, because of its cheerful, musical notes. The characteristic call is a long-drawn whistle dropping suddenly in pitch at the end: 'weeeeeeee-tu'. There are variants, typically a whistled phrase again ending in the final dropped note. This is a bird of the middle canopy, and is easily camouflaged in the foliage, being green above and yellow below, with prominent white and yellow bars in the black wing. Common in mangrove and woodland behind the beach, it can be found in any open woodland and gardens, especially in the lowlands, and it is also common in the Cibodas gardens. It ranges from India and China south to Java.

The leafbirds are slightly larger green birds of the forest canopy. Of the two species in Java, the BLUE-WINGED LEAF-BIRD *Chloropsis cochinchinensis* (18 cm) is quite common in forest margins, from sea level to about 1 500 m. As the name implies, the green plumage blends with the leaves, but the male has a black throat bordered by yellow, and both sexes have blue primaries and outer tail feathers. Although not as vocal as ioras, leafbirds have musical songs, but they are not very distinctive and are therefore often overlooked.

# Bulbuls

BULBULS are medium-sized birds of the middle storey, quite noisy, and often found in pairs or small parties. The plumage is soft and sometimes quite fluffy. While most species are rather dull, some are quite colourful, and some have short, attractive songs. Different species occur in open country and in the forest, and while the latter group may be difficult to identify, the former are familiar to many, especially as they are very popular cage-birds. There are 11 species in Java.

## SOOTY-HEADED BULBUL
*Pycnonotus aurigaster* (20 cm)
**Kutilang**                                          Plate 15

The common open-country bulbul of Java, it comes into gardens, whistling cheerfully from the tree-tops, or even from perches on town buildings. It can be seen in central Jakarta, and it occurs in both the plains and the hills. The loud and clear calls are musical but never merge into phrases to form a song. This is perhaps the commonest cage-bird in Java.

Among the cage-birds, the most expensive bulbul is the STRAW-HEADED *Pycnonotus zeylanicus* (29 cm), popular for its loud musical song, though this is always delivered at full volume and rather lacking in variation. Its large size and orange–yellow crown are distinctive, while the body is olive-coloured and streaked. This bird lives in dense vegetation in wet areas, but its popularity and price have now reduced it to being a very rare bird in Java. If steps are not taken to control the capture of wild birds, it could become scarce in Sumatra and Kalimantan as well. It is not known from Bali.

## YELLOW-VENTED BULBUL
*Pycnonotus goiavier* (20 cm)
**Terucuk**                                                    Plate 15

This is the common open-country bulbul of Sumatra and Kalimantan, but in Java it rather shuns the neighbourhood of man;
it occurs in similar terrain as the Sooty-headed, and is occasionally seen around Jakarta. The plumage is especially dowdy,
though the yellow under tail coverts and the whitish face with
a black stripe through the eye are distinctive. The brief song is
both unmusical and lacking in variety, without the rich notes
of some other bulbuls. It is common from southern Thailand
and the Philippines to Java.

## ORANGE-SPOTTED BULBUL
*Pycnonotus bimaculatus* (20 cm)
**Cica Rante**                                                 Plate 15

This typical open-country bulbul of the mountains is an endemic of Sumatra, Java, and Bali, common in all wooded areas
and forest margins above about 1 500 m. Olive and brown in
colour, paler on the belly, it also has yellow under tail coverts.
The head is quite different, however, being dark with prominent yellow ear coverts and orange spots between the bill and
eyes. This noisy bird gives some quite striking calls.

## BLACK-HEADED BULBUL
*Pycnonotus atriceps* (18 cm)
**Kuricang**                                                   Plate 15

Among the half dozen forest bulbuls, this is perhaps the easiest
to identify. Plate 15 illustrates the neat, colourful plumage.
Look especially for the yellow in the wing and the yellow tail
tip lined subterminally with black. The distinctive song consists mostly of single musical notes strung together to form

phrases in a regular pattern. This song can be heard regularly in the mornings around the forest margins and wilder wooded areas of the lowlands and lower hills. This species ranges from Assam and Vietnam south to Java.

Sometimes seen together with the Black-headed Bulbul is the similar RUBY-THROATED BULBUL *Pycnonotus dispar*. It differs in having a black crest, and lacks the prominent yellow in the wings and tail, but the most distinctive feature is the lovely brick-red patch on the throat.

# Drongos

DRONGOS are slender, all-black or dark grey birds with long and distinctively shaped tails. Two species in Java have very long 'racket' tails. The calls are varied, difficult to describe but once known, they are recognized as being typically drongo. Some of the calls are quite musical, and often one may be mystified by a new forest call only to find it is another variation from a drongo. Drongos are quite fierce, and are not afraid to drive off other birds twice their size.

## BLACK DRONGO
*Dicrurus macrocercus* (28 cm)
**Srigunting**                                    Plate 16

This common open-country drongo, a bird with a moderately long, deeply forked tail, often perches prominently on trees or telegraph wires. From this vantage point, it flies forth to catch insects in the air or even snatches them from the ground like a kingfisher might, and it is capable of skilful acrobatics. It is especially fond of flying termites, hawking them around the smoke of a fire. The usual call notes are quite harsh. This drongo ranges widely from Iran eastwards to China, with a closely related species in Africa, but it is absent in the equatorial zone, so that Java is a remote outpost.

The ASHY DRONGO *Dicrurus leucophaeus* (28 cm) replaces it in the forests, especially in the hills; it differs in being dark grey instead of black, and in having a less deeply forked tail. In the savanna woodlands of the east, such as Bali Barat, as well as on some off-shore islands, an observer should look for the SPANGLED DRONGO *Dicrurus hottentottus*, which is like the Black Drongo except that the tips of the tail are rounded and strongly up-curled.

## LESSER RACKET-TAILED DRONGO
*Dicrurus remifer* (27 cm)                                      Plate 16

The length of 27 cm does not include the enormously extended outer tail feathers or rackets, which themselves can be a further 50 cm. However, the rackets are sometimes missing; in such cases, this drongo is not very easy to distinguish from other black drongos. It occurs in montane forests, and is replaced in lowland forests by the very similar but larger GREATER RACKET-TAILED DRONGO *Dicrurus paradiseus* (body: 33 cm). Both birds have a wide range of musical notes. Only the latter bird occurs in Bali.

# Orioles

## BLACK-NAPED ORIOLE
*Oriolus chinensis* (27 cm)
**Kapodang**                                                    Plate 16

Most orioles have golden yellow plumage, though this species has a black stripe through the face and round the back of the neck, and also black primaries and central tail feathers. This fruit-eating bird of the canopy in open woodland and gardens is noted for its brief but rich, fluty warbled call notes described quite well by its name: 'o-ri-ol-lo'. Although it is not very conspicuous in the tree-tops, despite its bright colour, it nevertheless seems to have suffered badly from catapults, airguns, and the bird trade, and is not nearly as common as it used to be. Around Jakarta, one is likely to hear it regularly only in the zoo grounds. It is widely distributed throughout South and South-East Asia, through the Sundas as far east as Flores, though it seems to be extremely rare in Kalimantan.

In the mountains, one may be lucky to see the rather rare BLACK-AND-CRIMSON ORIOLE *Oriolus cruentus* (23 cm), which is mainly black, with light bluish feet and bill. On the breast and in the wing coverts, the male has dark crimson-red patches, so dark that in poor light, the bird may appear all black. The call is reputed to be a whining, nasal note.

# Crows

ALTHOUGH crows are well known as rather large, harsh-voiced, black scavengers of open countryside, it is not generally realized that the family includes a variety of jays and magpies, some with colourful plumage and melodious calls. However, the family is rather poorly represented in Java.

LARGE-BILLED CROW
*Corvus macrorhynchos* (50 cm)
**Gagak**                                                    Plate 16

This heavy, black bird with thick bill can be found in any type of open country, often in twos or threes, calling with single harsh 'kar' notes. Omnivorous, they feed on almost anything, from carrion to bird's eggs. Its nest is a large, bulky affair placed in the tops of tall trees. Formerly common, it has now become rare and local, probably having fallen victim to the ubiquitous catapult and airgun.

However, it can easily be confused with a second species, the smaller SLENDER-BILLED CROW *Corvus enca* (45 cm), which is primarily a forest bird though ranging out into wood-land margins. This crow can be distinguished by its more slender bill, shallower wing-beats, and higher-pitched voice. It often occurs in small flocks.

## RACKET-TAILED TREEPIE
*Crypsirina temia* (33 cm)                                     Plate 17

The length of this bird includes some 18 cm of long tail, which is broadened at the tip. Being wholly black with a glossy sheen, the treepie is rather drongo-like, but its bill is thicker and the tail is not forked. Unlike a drongo, however, it prefers to skulk about the middle storey of trees or bushes in lightly wooded areas, rather after the fashion of a malkoha, with occasional short flights to the next bush. It might pass unnoticed unless the harsh double notes attract attention. It is rather uncommon, though it may be seen near the south coast, and occasionally in scrubby country in the hills around Bogor.

## SHORT-TAILED GREEN MAGPIE
*Cissa thalassina* (32 cm)                                     Plate 17

One of the delights of bird-watching in open montane forests is to chance upon a party of these attractive birds. The green plumage, yellower on the under-parts, with red bill and legs, chestnut primaries, a black band through the eye, and a rather long graduated tail, is unmistakable. Despite this bright plum-age, they are quite difficult to see, even in bird waves, for they skulk about the thick vegetation of the middle and lower storeys, making only occasional brief flights, but their loud metallic calls and fairly melodious four-note song soon give them away. It is not found in Bali.

# Tits and Nuthatches

## GREAT TIT
*Parus major* (13 cm)
**Gelatik Batu**                                    Plate 17

This cheerful and familiar tree bird of lightly wooded country
and gardens ranges from North-west Europe to China and
south to Java and Flores. However, it is very rare in Malaysia,
Kalimantan, and southern Sumatra. In Java, it is a common
bird, and its varied 'pi-chu, pi-chu' calls can often be heard,
even in Jakarta gardens. Its black and white face and black
ventral bib are distinctive, though the rest of the plumage is
grey or greyish-white, lacking the bright yellow under-parts
and greenish upper-parts of its European cousins.

The PYGMY TIT *Psaltria exilis* (8 cm) is a tiny, rather plain
bird that is known only from the mountains of West and Central
Java. Its nearest relative lives in western North America! It is
brownish-grey, paler below, with a short bill and comparatively long tail. It is quite commonly found in pairs or small
active parties in the trees of the Cibodas gardens and adjacent
forest edges.

## BLUE NUTHATCH
*Sitta azurea* (13 cm)                              Plate 17

Small, compact, short-tailed birds, nuthatches behave rather
like woodpeckers, clambering about the trunks and branches
of woodland trees. Unique to nuthatches is their ability to also
climb vertically down the trunk, facing down. This species'
plumage is distinctive, with its black crown and belly, dark and
light blue upper-parts, white throat and breast, and pale blue bill.
This nuthatch, a common montane resident of Java, Sumatra,
and the Malay Peninsula, often occurs in small parties, especially in bird waves.

In lowland forests, it is replaced by the VELVET-FRONTED NUTHATCH *Sitta frontalis*, which differs in having a red bill, all whitish under-parts, and all blue upper-parts, black on the forecrown. No nuthatches occur on Bali.

# Babblers

BABBLERS, the bane of the casual bird-watcher or beginner faced by a host of 'small brown birds' in dense vegetation, are often difficult to see—or to describe adequately when seen—though nearly all have distinctive calls or songs which are a prime guide to identification. These songs, however, require much practice and experience to learn. While some species generally seem to live in pairs, more commonly they will be encountered in small family parties, scurrying about in the undergrowth or lower storey, rarely offering a view of more than a few seconds, but calling constantly to each other. Imitating their calls often serves to attract the inquisitive birds closer to the impostor. Java has a total of 19 species, of which 7 are endemic, but Bali has only 3 species. Only those which can be most readily seen or identified will be described here.

## HORSFIELD'S BABBLER
*Trichastoma sepiarium* (15 cm)
**Kancilan Sunda** Plate 18

This species, a member of the most difficult group, the jungle babblers, is a very common bird of thickets on forest margins or in dense woodland and wooded villages, and its song can often be heard in the public gardens at Bogor and, sometimes, Cibodas. The forest along the back of remote beaches is also a favourite habitat. Little can be said about the plumage, which is greyish to olive-brown, paler below, but notice should be taken of the short tail and rather thick bill. In contrast to the plumage, the song is easy to learn and can be heard every morning, and often at dusk as well. Unusually for a babbler, the song of this species is monotonous and repetitive, consisting of thrush-like notes often in threes strung together as 'chup, chwee, chwee', often interspersed with a single rising disyllabic 'hooeet', which is an accompaniment from the female (many babblers are noted for duetting between pairs).

Common in primary and secondary forest is the ground-frequenting BLACK-CAPPED BABBLER *Pellorneum capistratum* (18 cm). It is a chestnut-brown bird with a black stripe on the crown, and a rather short tail. Its simple call, 'pi-peee', with the second note lower, can be heard everywhere in suitable habitat.

## CHESTNUT-BACKED SCIMITAR BABBLER
*Pomatorhinus montanus* (20 cm)
**Burung Angklung** or **Burung Kopi-kopi**　　　Plate 18

Unlike many babblers, this bird is quite conspicuous and easy to identify. It is a common and noisy bird of montane forests, nearly always encountered in small flocks and always present in bird waves. The long, decurved bill and rather long tail are distinctive. It can be readily identified from the neat plumage as shown in Plate 18. The calls are loud and quite varied and mellow, though sometimes rather repetitive. Its range extends through Sumatra, Kalimantan, and Malaysia.

## PYGMY WREN-BABBLER
*Pnoepyga pusilla* (9 cm)　　　Plate 18

Most wren-babblers are small, round, and short-tailed, but this one is a tail-less little ball. It is brown all over, with pale underparts marked with black crescentic scales. A common bird of montane forests, it lives on the ground in dense undergrowth.

It is not found in Bali. It is not always difficult to see, but its voice, remarkably loud for so small a bird, can be heard everywhere: it consists of three sharp whistles uttered with long intervals between each, each note at a lower pitch.

## PEARL-CHEEKED BABBLER
*Stachyris melanothorax* (13 cm)                                      Plate 18

The Pearl-cheeked Babbler is a representative of the tree babblers, active and lively birds of the lower and middle storeys, often seen in parties. Many of them have some distinctive features of plumage to assist in identification, on the rare occasions one can get a good view, and the songs are distinctive. This small babbler lives in thickets in secondary growth and forest, and is a common endemic to Java and Bali. The call consists of short bursts of a rapid trilled note, uttered at different speeds but usually just too fast to enable a clear imitation. Such an imitation is always worth an attempt, however, as generally the birds will approach to see the intruder. Incidentally, this song is identical to that of the CHESTNUT-WINGED BABBLER *Stachyris erythroptera* of Sumatra and Kalimantan.

The WHITE-BREASTED BABBLER *Stachyris grammiceps* (15 cm) is a poorly known and possibly now very rare endemic, as it inhabits forests of the lowlands of West Java. There are few recent records, though it is perhaps rather inconspicuous. It is a beautiful chestnut-brown above, with a white breast and belly, and grey head and flanks.

The WHITE-COLLARED BABBLER *Stachyris thoracica* (18 cm) is a larger and more distinctive endemic, although uncommon and difficult to observe. It occurs in small parties in the forest undergrowth in the hills up to 1 600 m. It is dark chestnut, with a broad white band across the breast, and dark brown and grey on the head. The song is said to consist of about five notes down the scale, ending in a flourish; it also gives a rather harsh, rattling note.

The JAVAN TIT-BABBLER *Macronous flavicollis* (14 cm) is another Javan endemic, though very similar species occur in Palawan and Vietnam. It occurs in small parties in the undergrowth of forest edges and scrub in the lowlands, and is not very common. It is pale brown, chestnut on the wings, grey on the head, rich tawny on the breast with scarcely visible black streaks, and buffy white on the belly. Calls include rather frog-like chattering and a regular 'tonk, tonk, tonk' call. However, it can be confused with a more widespread Sundanese species, the STRIPED TIT-BABBLER *Macronous gularis*, which has a similar call; the blackish streaks on the under-parts are much more pronounced.

## RED-FRONTED LAUGHING-THRUSH
*Garrulax rufifrons* (27 cm)
**Burung Kuda-kuda**                               Plate 18

Endemic to Java, this is the sole representative of a large group centred on the Indo-Chinese region. Laughing-thrushes are comparatively large, floppy, rather long-tailed babblers, delightful in voice and mannerisms. They move in pairs or small parties through the undergrowth with effortless hops and short flights. Some species have loud, raucous calls that climax in several birds producing an almost maniacal laugh, but the Red-fronted is more subdued. Its usual song is a series of deliberate whistles on one pitch, uttered at a rate of about three per second for 5–10 seconds, which the Javanese liken to a horse-like giggle. It is quite common in montane forests, and is often trapped for the cage-bird trade.

## CHESTNUT-FRONTED SHRIKE-BABBLER
*Pteruthius aenobarbus* (12 cm)
**Ciu Kecil**                                        Plate 18

The shrike-babblers are colourful babblers of the middle and upper storeys of montane forest. This species is slightly smaller than the Common Iora, but with its double white wing bar (rufous in the female), it is rather similar. Look for the chestnut patch around the bill and on the throat of the male. The female is paler, almost lacking the yellow on the belly and has only a rufous patch on the forehead. They can most readily be seen in bird waves.

There is another, larger species that occurs in the same habitat, the WHITE-BROWED SHRIKE-BABBLER *Pteruthius flaviscapis* (14 cm). The plumage is quite different, being white below and blackish above, with a prominent white eyebrow and an orange–yellow patch on the wing. The female is brownish. No shrike-babblers are found in Bali.

The JAVAN FULVETTA *Alcippe pyrrhoptera* (14 cm) is another endemic to the western half of Java, where it is common in hill forests, living in the middle storey in small active parties. It is rich brown above, paler below, with a rather distinctive square-shaped head. The song is a rapid tinkling cadence.

The last endemic member of the babbler family is the SPOTTED CROCIAS *Crocias albonotatus* (20 cm), a canopy bird of montane forests in West Java. It is not easily seen, unless a mixed feeding party or bird wave brings it down to closer range. A rather long and graduated tail, tipped white, is distinctive. The top of the head is black and the back is chestnut, strongly spotted and streaked with white. The under-parts are clean white.

## Thrushes

THE thrushes form a very large family encompassing the true thrushes, chats, forktails, and robins. Some are renowned songsters. Mainly insectivorous birds of woodland, open country, and sometimes rushing rivers, they move with a hopping gait.

## WHITE-BROWED SHORTWING
*Brachypteryx montana* (15 cm)
**Cingcoang Alis Putih**                                    Plate 19

Shortwings are skulking birds of montane forest, having short wings and short tails. The male is all blue, while the female is blue and bright rufous. There is a concealed silvery white streak over the eye of the male. Strongly territorial, it will sometimes make a circuit of its territory, singing once from each corner. The song consists of about three strongly emphasized and well-spaced notes that break into a short, fairly rich warble.

Two species occur in Java and both can be heard commonly at Cibodas. The LESSER SHORTWING *Brachypteryx leucophrys* (12 cm) is brown in both sexes, with a short, white eyebrow in the male. The brief warble, more jangling than that of the White-browed species, begins slowly but lacks the emphatically spaced notes. There is a quite distinct altitudinal separation between the habitats of the two species, the Lesser occurring below 1 600 m (it can be heard behind the Cibodas rest-house) while the White-browed occurs most commonly from 1 600 to 3 000 m. They overlap between 1 400 and 1 600 m. Only the Lesser Shortwing occurs in Bali.

## MAGPIE ROBIN
*Copsychus saularis* (22 cm)
**Kucica**                                                  Plate 19

Most people know the black and white Magpie Robin, a favourite cage-bird and a common participant in song contests. Common in gardens and villages, or lightly wooded open country, its popularity seems to have reduced its numbers considerably in the more densely populated areas. A bird of the middle and lower storeys sometimes coming down to the ground, it usually cocks its tail as it alights. Magpie Robins in Bali and East Java differ in having a black belly. The song, quite a sweet warble, lacks the rich fluty tones of a related species, the WHITE-RUMPED SHAMA *Copsychus malabaricus* (27 cm). This black bird with rufous belly and prominent white rump and a longer tail occurs in the deep forest of the lowlands and is thus now quite localized in Java. Presumably, many of those seen in the bird markets originate from Sumatra or Kalimantan. The Magpie Robin has a wide distribution from Pakistan to South China, south to Java and the Philippines (but not Sulawesi), and the Shama's range is nearly as wide.

## WHITE-CROWNED FORKTAIL
*Enicurus leschenaulti* (25 cm)
**Meninting Besar**                                        Plate 19

Forktails are typically birds of fast-running streams, where their long, forked tails are apparently designed to provide balance when they perch on rocks. Characteristic of many species in this environment, the calls consist of single piercing whistles, which can carry above the noise of the water. This species occurs along forested streams in both the lowlands and the mountains, though it is commoner in the mountains where this habitat is less disturbed. However, it is not very easily spotted as it is shy, and generally the call attracts attention to the bird as it dashes past to a perch out of view. The black and white plumage might appear to be diagnostic, but note should also be taken of the LESSER FORKTAIL *Enicurus velatus* (16 cm), a smaller bird, with shorter tail, that is a montane endemic of Java and Sumatra (though absent from Bali). In this bird, the upper-parts are grey with a chestnut or dull chocolate crown, and both the breast and belly are white. Both species occur at Cibodas, though the latter is usually at higher elevations. The White-crowned Forktail ranges from the western Himalayas and South China south to Java.

## PIED BUSHCHAT
*Saxicola caprata* (14 cm)
**Kucica Batu**                                    Plate 19

This bird of open country perches on low bushes and stones
and sometimes tree-tops, telegraph wires, and roof-tops. It has
a habit of frequently flicking its tail. It is an insect-eater, catching its prey on the ground and sometimes in the air. The male's
black plumage with white rump and wing patch is distinctive.
The female is dark brown, slightly streaked, and without the
white wing bar, but with a distinctive rusty rump. This species
is usually seen in pairs. The song is a short though somewhat
nondescript warble, while the normal call note is quite harsh.
The Pied Bushchat ranges from the Middle East to the Philippines, and from Java to Sulawesi and Papua New Guinea. In
Java it can be found from sea level to intermontane plateaux
such as the Ijen.

## SUNDA WHISTLING THRUSH
*Myophoneus glaucinus* (24 cm)
**Tiung Batu**                                    Plate 19

Rather large, usually dark blue birds, whistling thrushes are
sometimes quite noisy and conspicuous denizens of the lower
storey of dark montane forest and moss forest, often found near
streams. The male of this species is dark purplish-blue, slightly
glossy above, and unmarked, while the female is dark brown.
The black bill is a useful character in identification. The tail is
frequently fanned when the bird hops on the ground. The call
is a trisyllabic, rather squirrel-like chortle, but the song is not
unmusical.

Confined to the mountains of Java, Bali, Sumatra, and Kalimantan, it is quite common in Java, but one should take care
not to confuse it with the very similar BLUE WHISTLING
THRUSH *Myophoneus caeruleus* (32 cm). This larger bird is

identified by its yellow bill. It is shy and uncommon, usually seen near mountain streams, where it gives a shrill whistle like that of a forktail.

These blue thrushes could be confused with the JAVAN COCHOA *Cochoa azurea* (25 cm), a rare endemic of the West Java mountains. The male is purplish-black, shading to pale blue on the crown and wings, and the female is browner. Although quite tame, it moves silently around the lower storey and is easily overlooked unless the song is heard. This thrush's song has been likened to the notes of a bamboo flute.

## ORANGE-HEADED THRUSH
*Zoothera citrina* (22 cm)
**Burung Anis Merah**                                    Plate 19

There are several thrushes in the mountains of Java, none of them very common. The most distinctive and beautiful is the Orange-headed, for which the illustration in Plate 19 is sufficient description. A bird of the ground and lower storey of mainly hill forests, it ranges from the Himalayas south to Sumatra, Kalimantan, and Java. Despite its colourful plumage, it is not easily spotted, being a shy bird of the forest floor, but it has a loud thrush song. It is popular in the cage-bird trade and has become uncommon. One or two still live in the Bogor Botanical Gardens though it is expected that they will soon become extinct there.

# Warblers

THE warblers are thin-billed insect-eaters that hunt actively in the foliage. Most are small and generally brown, or yellow and green, but there is a wide variety of forms. Many have distinctive songs.

## FLY-EATER
*Gerygone sulphurea* (9 cm)
**Burung Remetuk**                                    Plate 20

The Fly-eater is best known for its song, as it is generally
hidden in the canopy. Especially common in mangroves, it
occurs also in a wide range of wooded habitats, including
casuarinas, rubber, and sometimes secondary growth and
forest. It can even be heard by tree-lined canals in downtown
Jakarta. The song, one of the characteristic sounds of Pulau
Dua through the heat of the day, consists of repeated phrases
of about three to five notes on a descending scale, wheezy and
uttered as if with effort. Once learnt, by observing the bird
give this call, it is easily remembered. The rounded head and
rather large bill is unusual for a warbler.

## SUNDA WARBLER
*Seicercus grammiceps* (10 cm)                        Plate 20

This brightly coloured little warbler is common in montane
forest and forest edge in Sumatra, Java, and Bali. Mostly in the
middle canopy, it descends to the undergrowth and commonly
joins bird waves (mixed species flocks). The call consists of
the typical, thin shrill notes of small canopy warblers, and is
not very distinctive. In the same habitat can be found the
MOUNTAIN LEAF WARBLER *Phylloscopus trivirgatus*
(11 cm), greenish above and yellowish beneath, with longi-
tudinal bands of black and yellow on the head (black crown
and eye-stripes separated by yellow). It is Java's sole resident
representative of a large northern, migratory genus, many of
which are notoriously difficult to identify.

## STRIATED WARBLER
*Megalurus palustris* (25 cm)
**Bejuwit** Plate 20

This long-tailed brown warbler of open habitats is easily identified by its large size, graduated tail, and boldly streaked brown plumage. The harsh, unvarying song always gives it away and is best described by its Sundanese name of 'Cek, Cek, Kored', uttered from a low perch. It is a very common bird of open country, from wet coastal grasslands to open scrub and tea estates in the mountains. In the Puncak, the name of 'tea-lurk' seems to aptly describe its behaviour. It is a bird of monsoon Asia, with outposts in the Philippines and Java, and a related species replaces it in Sulawesi, Nusa Tenggara, and Ceram.

## ASHY TAILORBIRD
*Orthotomus ruficeps* (11 cm)
**Cinemen** Plate 20

There are two small, long-tailed warblers that are active and at times noisy in town gardens, a tailorbird and a prinia. Tailorbirds are rather skulking little birds of the undergrowth and scrub, so named for their remarkable woven domed nests. The Ashy Tailorbird is generally the most common, and can be found in gardens from Jakarta to Cibodas. It is dull greenish above, more grey below, and identified by the rufous crown and cheeks. The tail is quite long, graduated, and greenish in colour. The rather shrill call is usually disyllabic.

Another open-country species is the LONG-TAILED TAILORBIRD *Orthotomus sutorius* (12 cm), which has a similar distribution, though it is absent in Bali. Instead of having a rufous head, it has chestnut confined to the fore-crown. In breeding plumage, the tail of the male is elongated by a further 3 or 4 cm. The call is almost monosyllabic, and may be repeated endlessly for minutes at a time.

In the mountains, look out for the MOUNTAIN TAILORBIRD *Orthotomus cuculatus* which is distinguished by its longer bill, long white eyebrow, and bright yellow belly. It differs from other tailorbirds in having a varied musical repertoire, and also by not having woven nests.

## BAR-WINGED PRINIA
*Prinia familiaris* (13 cm)
**Perenjak Sayak Garis** Plate 20

Prinias differ from tailorbirds in their generally longer tails, which are also slightly graduated, and plain brownish to greenish plumage. This species is greenish-brown, with a pronounced double yellowish-white wing bar. As implied by its scientific name, it is indeed a familiar bird, in gardens and thickets in open country. As much at home in the garden of a Jakarta hotel as it is in the Cibodas Botanical Gardens, it is much given to sudden cheerful call notes, loud for so small a bird. It is endemic to Sumatra, Java, and Bali.

Another little member of the warbler family is the JAVAN TESIA *Tesia superciliaris* (7 cm), which is endemic to the mountains of western Java. It is a tiny, tail-less bird that is active in the ground vegetation and easy to observe. Olive-brown above and light grey below, it has a dark crown and long whitish eye-stripe. Usually seen in pairs, this warbler is very vocal, with a variety of loud, clear notes and short metallic trills.

A very poorly known bird is the JAVAN BUSH WARBLER *Bradypterus montis* (13 cm), found in the higher mountains of East Java and Timor. It creeps rather stealthily about the ground shrubs in the moss forest zone or higher, but has some loud call notes. It is rather nondescript and brownish, darker above and paler below, with a whitish eye-stripe and rather long tail. (Some authors treat this species as a race of the Russet Bush Warbler *B. seebohmi* of continental South-East Asia.)

# Flycatchers

THIS is quite a variable and sometimes colourful family of insect-eaters. Some forage after the manner of warblers, while others will fly out from a perch to catch insects in mid-air. Some are quite sluggish but many have a habit of frequently flicking their tails when perched. They are mostly found in the middle canopy. A total of 23 species (including the whistlers) are found in Java, and only a few characteristic ones are described here.

## INDIGO FLYCATCHER
*Muscicapa indigo* (14 cm)                                    Plate 21

The Indigo Flycatcher is one of a group of blue flycatchers, all of which are rather alike and, thus, sometimes difficult to identify. The songs of all of them consist of a rather thin warble, typically flycatcher but rarely distinctive. This bird is blue all over except for a paler forehead, white under tail coverts, and blackish lores. It occurs in the montane forest of Sumatra and Java, and is not very conspicuous except when it joins the excitement of a bird wave.

At lower elevations may be found the HILL BLUE FLY-CATCHER *Cyornis banyumas* (15 cm), which is dark blue above, shining blue on the forehead, and bright rufous on the breast grading to a pale belly. The female of this species is brown. Neither species occurs in Bali.

## SNOWY-BROWED FLYCATCHER
*Ficedula hyperythra* (11 cm)                                Plate 21

This common little flycatcher is generally seen in pairs in the lower and middle storeys of montane forests. Although not obtrusive, it is inquisitive and quite tame. The male is dark blue above and rufous below, with a pronounced white eyebrow, and the female is buffy-brown with a paler eyebrow.

## PIED FANTAIL
*Rhipidura javanica* (18 cm)
**Kipasan**                                                  Plate 21

Fantails are large flycatchers with long, broad and slightly graduated tails. The Pied Fantail, a familiar bird of scrub, especially in damp places, is common also in mangroves and sometimes town gardens. Brown above, with white tips to the tail feathers, it is readily identified by its white under-parts and broad black breast band. As its name implies, it frequently fans its tail. The song is a short warble, loud and distinctive though difficult to describe. It is common from Thailand to the Philippines, south to Sumatra, Kalimantan, Java, and Bali.

There are also two endemic fantails on Java. One is the RED-TAILED FANTAIL *Rhipidura phoenicura* (17 cm) that is common in the forests of mountain areas. This conspicuous bird of the lower storey is almost invariably present in bird waves. It is dark below with a white breast band, and the belly and constantly flicked tail are bright rufous.

The WHITE-BELLIED FANTAIL *Rhipidura euryura* (18 cm) is also a Javan endemic, although some authors treat it as a race of the more widespread SPOTTED FANTAIL *R. perlata*. It is a blue–black bird with a pronounced short, white supercilium, white belly, and white outer tail feathers; it lacks the heavy white spotting on the breast that characterizes the Spotted Fantail in Sumatra. It is quite common in hill forest up to about 2 000 m, and often joins mixed feeding flocks. In some ways it is rather sluggish and does not adopt the characteristic horizontal pose, swinging laterally and fanning its tail, that is typical of fantails.

## ASIAN PARADISE FLYCATCHER
*Terpsiphone paradisi* (20 cm, plus 23 cm of elongated tail plumes in the male).
**Burung Sriwang**                                    Plate 21

This is the most striking flycatcher of all. Snowy white, with a glossy black head and throat, a small crest, and blue bill and eye-ring, the male has immense, long white tail plumes. The female is rufous with a black head, paler under-parts, and a moderately long tail. There are two colour phases, however, and some males have the same colour as the female but with a longer tail. Found from the Himalayas and China south to Java and Flores, it is a bird of lowland forest and is therefore now rather rare in Java. It lives in the middle storey, where it is not very conspicuous despite the startling plumage of the white phase male, though the loud 'wit-wit-wit-wit' call often gives it away.

# Wagtails and Pipits

## YELLOW WAGTAIL
*Motacilla flava* (18 cm)
**Entut Kerbau**                                       Plate 22

The Yellow Wagtail, a common winter visitor to the ricefields of the north coast of Java, occurs in small parties feeding mainly in short grass. Using the moderately long tail as a balance, it walks or makes short runs on the ground. The bird is olive-green above and yellowish below, with white outer tail feathers and pale wing bars. When flushed or in flight, it utters freely a wheezy 'tseep'. The slightly larger and longer-tailed GREY WAGTAIL *Motacilla cinerea* is usually solitary, found more often in the mountains, especially near rivers and roads. The upper-parts are grey, not olive, and the throat is whitish; there is also some white on the wing. However, immature birds are not very distinctively coloured, and the habitat and length of the tail are the safest features.

Pipits are brown-streaked, mainly terrestrial birds of open country, rather similar to larks. The RICHARD'S PIPIT *Anthus novaeseelandiae* (18 cm) occurs anywhere in open country but is not common. It differs from a lark in being more slender, with a longer tail and rather long legs. The plumage is paler than the Bushlark's, with no rufous in the wings. This species feeds on short grass or among crops, often making little runs to catch an insect.

# Wood-Swallows

WHITE-BREASTED WOOD-SWALLOW
*Artamus leucorhynchus* (18 cm)
**Kekep**                                               Plate 22

The most characteristic features of wood-swallows are their triangular wings and soaring flight. This common bird of open country is found from open spaces in central Jakarta up to the gardens of Cibodas. They perch in pairs or in little family groups, huddled together in a line on branches, cables, or the edges of buildings, periodically taking off to soar in circles in search of insects and then returning to perch. It has a rather squat shape, short square tail, and stout bill, and is slaty grey above and white below, with a pale rump. A member of a mainly Australian family, this species ranges from Australia and Fiji to Kalimantan, Sumatra, and the Andaman Islands. A similar species replaces it on mainland Asia.

# Shrikes

## LONG-TAILED SHRIKE
*Lanius schach* (25 cm)
**Bentet Biasa**                                        Plate 22

Shrikes are medium–large birds with strong heads and hooked
bills, and rather long tails. They feed by pouncing on insects,
and even small birds, from an exposed perch, sometimes
impaling surplus food on thorns as a sort of larder. Pale grey
above and whitish below, with black forehead, wing feathers,
and tail, and light rufous back, wings, and flanks, it has a broad
black mask through the eye. While the call notes are harsh, it
does have quite a sweet, brief warbling song. This species is
quite common in Java, and it, too, can be found from open
spaces in central Jakarta up to the gardens of Cibodas. Its range
extends from India and China to Papua New Guinea.

# Starlings and Mynas

MEDIUM-SIZED birds, with short tails and strong, pointed bills,
starlings are gregarious and mainly arboreal, though some
species feed mostly on the ground, often around cattle. Their
flight is swift and direct, though this is less so in the mynas,
which have more rounded wings.

## PHILIPPINE GLOSSY STARLING
*Aplonis panayensis* (20 cm)                           Plate 22

The Philippine Glossy Starling is a sleek, black bird with distinctive red eyes, although in good light, the plumage is seen
to be glossy green. It lives in small parties in open countryside,
even ranging into the hills, though it is most abundant in
coastal coconut groves. It breeds in open communities in holes
in trees, and its single, shrill call notes produce quite a cacophony of sound at large evening roosts. It is more often seen in

eastern and southern Java, and seems to be very rare on the northern coastal plains.

## ASIAN PIED STARLING
*Sturnus contra* (24 cm)
**Jalak Suren**                                                    Plate 22

This starling, readily recognizable by its black and white plumage, and orange skin around the eye, is quite common in the open country of East Java, but has become rather scarce along the populous northern plains. Many starlings and mynas are caught by bird snarers who often use pinioned birds to attract flocks down to the ground. This and the WHITE-VENTED MYNA *Acridotheres javanicus* are both ground feeders. The latter bird is black and greyish-black, with a prominent white patch on the wing and under the tail, and yellow–orange bill, feet, and eye-ring. These two species have the same distribution and habitat, and indeed often occur together. The Pied Starling ranges from India and South China to Java, but favours the monsoon climates. Any small populations in the equatorial regions of Sumatra and Kalimantan will be feral, except perhaps in Lampung.

The GRACKLE or HILL MYNA *Gracula religiosa* (30 cm), such a favourite cage-bird with its ability to mimic human speech, has now become rare in Java. It is found in lowland forests. It is glossy black with a white patch in the wings, and small yellow pouches round the back of the head. A favourite pastime of some owners is to teach their birds to sing the Indonesian national anthem; sometimes the birds are also taught more salacious expressions!

## BALI WHITE MYNA
*Leucopsar rothschildi* (25 cm)
**Jalak Putih Bali**                                    Figure 6

The Bali Myna is one of Indonesia's most famous birds, and one of the world's rarest. Pure white, with white crest, blue facial skin, and black-tipped wings and tail, the wild population may number no more than about 50 birds, confined to the Bali Barat Reserve. It should not be confused with the BLACK-WINGED STARLING *Sturnus melanopterus*, which is also white but with wholly black wings and tail and yellow facial skin. Also an endemic, it occurs in both Bali and Java, but is more common in the eastern areas. A few are seen occasionally in the west, even around Jakarta. Both species are popular as cage-birds, and the high price of the Bali White Myna is undoubtedly the main factor in its decline. However, this beautiful bird breeds readily in captivity, and a programme is in progress to reimport surplus zoo stock from America and to release the captive-bred birds back into Bali Barat.

6. Bali White Myna
   *Leucopsar rothschildi*

# Sunbirds

SUNBIRDS might be considered as the oriental equivalent of the Neotropical hummingbirds, being tiny, active little nectar-feeders with slender, curved bills. They feed on both nectar and insects, and are always busy about flowering shrubs, though they do not have the capacity of prolonged hovering. The males have brilliant glossy plumage, though in the field this often looks black. Females, young, or males in moult have dull plumage and they are very difficult to identify. There are eight species in Java, and four of the very long billed spiderhunters.

## BROWN-THROATED SUNBIRD
*Anthreptes malacensis* (14 cm)
**Burung Madu**                                    Plate 23

Probably the commonest and most widespread sunbird, it occurs in open country, especially near the coast, favouring coconuts, but also in urban gardens, and extending inland to almost the altitude of Cibodas. The main features in the male are the brown throat and yellow belly, though one should look out for the similar RUBY-CHEEKED SUNBIRD *Anthreptes singalensis*, which has a rufous throat, in woodland and forest edges. The upper-parts are metallic green, grading to purple on the rump and wing coverts. By contrast, the female is olive-brown above and yellow below, a broad description that fits most female sunbirds.

There is also another rather similar open-country species, the OLIVE-BACKED SUNBIRD *Nectarinia jugularis* (11 cm). The male of this species has non-metallic olive upper-parts and a yellow belly but a metallic blue throat. The bill is finer, and both sexes have white tips to the tail feathers.

## KUHL'S SUNBIRD
*Aethopyga eximia* (13 cm)                                    Plate 23

This Javanese endemic sunbird is probably also the most spectacular. It is a common and fairly tame bird of montane forest edges; the author found it to be abundant in coffee plantations on the Ijen Plateau. The long, graduated tail plumes are present only in the male in breeding plumage, and many birds look more like the rather dowdy female, olive above and pale greyish below. In all plumages, however, fluffy white flanks are clearly seen when the wings are raised.

In the same habitat may also be found the rather similar SCARLET SUNBIRD *Aethopyga mystacalis* (11 cm), which has a crimson mantle and breast, metallic blue crown, and a long, purplish, graduated tail in the breeding male. Some authors treat it as another endemic. Neither species is found in Bali.

## LITTLE SPIDERHUNTER
*Arachnothera longirostra* (16 cm)
**Burung Jantung**                                           Plate 23

Spiderhunters are similar to sunbirds but have extremely long, decurved bills and mainly green plumage in both sexes. There are four species in Java, of which this is the commonest and also the most readily recognizable because of its yellow belly and greyish-white throat. A bird living mainly in the forest edge, it especially favours wild bananas and ginger. The flight is direct and swift, and it is not often possible to get good views of any spiderhunter. The single call note can be repeated with great monotony.

# Flowerpeckers

SIMILAR in many ways to the sunbirds, the flowerpeckers have shorter bills. They are extremely active, stocky little birds, the males usually very colourful though the females are, again, drab and difficult to identify. They feed in the tops of flowering trees and epiphytes, and have swift, erratic flight. There are seven species in Java.

### SCARLET-HEADED FLOWERPECKER
*Dicaeum trochileum* (8 cm)
**Burung Cabe** Plate 23

This common open-country flowerpecker is an endemic to Java, Bali, and Lombok, with a relic population in South Kalimantan, though it has recently been discovered also in South Lampung. It visits a wide range of habitats, from urban gardens up to Cibodas, but it does not often stay still long enough to give a good view as it flits restlessly about the canopy of flowering trees, generally in pairs, giving sharp little 'chit, chit' notes. The female is brown above and pale below, but it also has the crimson rump of the male.

In more densely forested habitats, one may find the ORANGE-BELLIED FLOWERPECKER *Dicaeum trigonostigma* (8 cm). It is grey to blue but with prominent orange–yellow belly, back, and rump, and a pale grey throat.

### JAVAN FIRE-BREASTED FLOWERPECKER
*Dicaeum sanguinolentum* (8 cm) Plate 23

This flowerpecker is also an endemic, being found only in the mountains of Java, Bali, and parts of Nusa Tenggara. In Cibodas, it overlaps with the Scarlet-headed, but is more confined to forest edges and even the forest itself. It is quite common, and is equally colourful. The male is dark blue above with a fire-red breast and creamy white belly, with a dark central line. As usual, the female is drab.

# White-eyes

## ORIENTAL WHITE-EYE
*Zosterops palpebrosus* (11 cm)
**Burung Kacamata**                                    Plate 23

Although not related to sunbirds and flowerpeckers, white-eyes are rather similar in their habitat preference and behaviour. Their taxonomy is complex, and the different species and races are quite difficult to separate in the field. They differ principally according to the shade and extent of yellow on the under-parts. For the amateur, it is sufficient to know that the common species in Java is the Oriental White-eye. It lives in loose parties that flit actively around the canopy of ornamental shade trees in the gardens and suburban avenues, constantly keeping in contact with sweet cheeping calls resembling those of baby chickens. The olive-green upper-parts and yellow under-parts grading to whitish on the belly is a plumage common to many birds, but the white ring around the eye is prominent and distinctive. It ranges from Jakarta up to Cibodas, and is abundant, for example, in the tree-lined avenues of suburban Bogor. It is replaced by other species in the higher mountains, in the mangroves, and on small off-shore islands.

In montane forests, the commonest white-eye might not be recognized immediately as it lacks the white eye-ring! This is the JAVAN GREY-FRONTED WHITE-EYE *Lophozosterops javanicus*, which is olive-green with a grey head. There may be some white on the face but this does not form a distinct eye-ring. This species occurs on mountains only in Java and Bali, in small parties which give a deeper, more metallic note than other white-eyes.

# Sparrows, Weavers, and Munias

THE last family to be dealt with in this book is the sparrows and the seed-eaters that plunder the ricefields, the so-called paddy or rice birds which farmers' children spend hours driving away from crops. They are small, rounded, gregarious birds with short, thick bills, sometimes quite colourful, and often popular as aviary birds because of their lively nature and sweet cheeping calls.

## TREE SPARROW
*Passer montanus* (15 cm)
**Burung Gereja**                                    Plate 24

Opportunists, the sparrows have followed man round the world and formed feral populations in towns and cities throughout Asia. This species ranges from North-west Europe to Java. It is assumed to be indigenous to Java, where it is common in towns and ricelands throughout, but is believed to have been introduced, perhaps on board ships, into other equatorial islands such as Sumatra, Borneo, Sulawesi, and the Philippines. On some of these islands, especially Borneo, it is quite scarce, but these birds are also rather overlooked by foreign ornithologists who are familiar with them, or with similar species, in their home countries. With its rather dowdy plumage, the sparrow is not a beautiful bird, but one must admire it for its adaptability and cheeky cheerfulness. They are not stupid birds, as anyone attempting to shoot them with a catapult will discover.

17. (a) Blue Nuthatch *Sitta azurea*. (b) Racket-tailed Treepie *Crypsirina temia*. (c) Short-tailed Green Magpie *Cissa thalassina*. (d) Great Tit *Parus major*.

18. (a) Chestnut-fronted Shrike-Babbler *Pteruthius aenobarbus*. (b) Pygmy Wren-Babbler *Pnoepyga pusilla*. (c) Chestnut-backed Scimitar Babbler *Pomatorhinus montanus*. (d) Red-fronted Laughing-Thrush *Garrulax rufifrons*. (e) Horsfield's Babbler *Trichastoma sepiarium*. (f) Pearl-cheeked Babbler *Stachyris melanothorax*.

19. (a) Pied Bushchat *Saxicola caprata*. (b) White-browed Shortwing *Brachypteryx montana*. (c) Sunda Whistling Thrush *Myophoneus glaucinus*. (d) Orange-headed Thrush *Zoothera citrina*. (e) White-crowned Forktail *Enicurus leschenaulti*. (f) Magpie Robin *Copsychus saularis*.

20. (a) Striated Warbler *Megalurus palustris*. (b) Sunda Warbler *Seicercus grammiceps*. (c) Ashy Tailorbird *Orthotomus ruficeps*. (d) Bar-winged Prinia *Prinia familiaris*. (e) Fly-eater *Gerygone sulphurea*.

21. (a) Asian Paradise Flycatcher *Terpsiphone paradisi*. (b) Snowy-browed Flycatcher *Ficedula hyperythra*. (c) Pied Fantail *Rhipidura javanica*. (d) Indigo Flycatcher *Muscicapa indigo*.

22. (a) Long-tailed Shrike *Lanius schach*. (b) Philippine Glossy Starling *Aplonis panayensis*. (c) Yellow Wagtail *Motacilla flava*. (d) White-breasted Wood-Swallow *Artamus leucorhynchus*. (e) Asian Pied Starling *Sturnus contra*.

23. (a) Brown-throated Sunbird *Anthreptes malacensis*. (b) Little Spiderhunter *Arachnothera longirostra*. (c) Kuhl's Sunbird *Aethopyga eximia*.
(d) Oriental White-eye *Zosterops palpebrosus*. (e) Javan Fire-breasted
Flowerpecker *Dicaeum sanguinolentum*. (f) Scarlet-headed Flowerpecker
*Dicaeum trochileum*.

24. (a) Tree Sparrow *Passer montanus*. (b) Streaked Weaver *Ploceus manyar*. (c) Java Sparrow *Padda oryzivora*. (d) Pin-tailed Parrotfinch *Erythrura prasina*. (e) White-headed Munia *Lonchura maja*.

## STREAKED WEAVER
*Ploceus manyar* (14 cm)
**Manyar**                                    Plate 24

Weavers are quite noisy birds, with cheerful, high-pitched chatters and wheezes, especially about the nesting colonies; they are colonial birds of the open country, particularly in ricefields. Their elaborately woven suspended nests in palms and rushes are bulky affairs, sometimes with a long entrance tunnel below; the shape of the nest is often a guide to identification. Of this mainly African genus, there are three species in Java, the limit of their range. Females and non-breeding males are dingy coloured, rather like female sparrows, and almost impossible to separate, but in the breeding season, the males become bright yellow about the head, with black facial masks. As Plate 24 shows, the Streaked Weaver is heavily streaked on both the back and breast, and this is a clue on both sexes. The breeding male has a yellow crown, and black head and throat. It is perhaps the commonest weaver, widespread in open country in the plains. The nests of this species have only a short entrance spout beneath.

The BAYA WEAVER *Ploceus philippinus* is similar, but the yellow on the crown is more extensive, with only a black facial mask, and unstreaked throat and breast. The nest has a long entrance spout. On the flat rice plains of the north coast, one should look also for the beautiful GOLDEN WEAVER *Ploceus hypoxanthus*, which has a golden head and under-parts, and a black mask and throat.

## JAVA SPARROW
*Padda oryzivora* (16 cm)
**Gelatik**                                    Plate 24

The Java Sparrow is the true 'paddy bird'. Originally endemic to Java and Bali, it has been introduced all over South-East

Asia where it can be found in a wide range of open habitats, including urban areas. Essentially a bird of riceland and open cultivation, in Java, however, it has become generally rather scarce, owing to its popularity as a cage-bird and perhaps as a rather prominent target for the bird-scarer's catapult. Even in Java it favours man-made environments, usually nesting about houses, or in hollows in trees. The adult is unmistakable with its prominent white cheeks surrounded by black, and thick red bill.

## WHITE-HEADED MUNIA
*Lonchura maja* (11 cm)
**Bondol Haji**                                          Plate 24

The munias are the smallest birds of the ricefields, though in fact they are found in a variety of open habitats. For example, this species is common from the scrub growth on Pulau Dua to the ricefields and vegetable plots in the mountains. Although seemingly tame, these gregarious, active little birds are quite wary and difficult to approach. The chestnut-brown plumage grading to white on the head is distinctive. They nest in low shrubs, paddy, grassy wilderness, trees and palms; sometimes two or even three species breed in one small tree as if it were a multi-storey apartment block.

The two other common species are the SCALY-BREASTED MUNIA *Lonchura punctulata*, which is brown above, becoming chestnut around the bill, and white below, heavily spotted or scaled; and the JAVAN WHITE-BELLIED MUNIA *Lonchura leucogastroides*, greyish-brown with a blackish breast and unspotted white belly.

## PIN-TAILED PARROTFINCH
*Erythrura prasina* (13 cm)
**Bondol Hijau**                                    Plate 24

Although related to the munias, this particularly colourful
little bird is much more secretive, living in well-wooded areas,
scrub, and bamboo, and never venturing far from this habitat
into adjacent open fields. It is not as gregarious as the munias,
and will suffer from the competition of a mixed aviary. The
male is extremely colourful with red elongated tail feathers.
These are lacking in the female, as are the patches of blue on
the head and red on the belly.

The TAWNY-BREASTED PARROTFINCH *Erythrura
hyperythra* occurs in the forests and bamboos of the mountains.
It appears to be a much rarer bird, but is perhaps overlooked.
The male is like a strongly coloured version of the female of the
Pin-tailed species, green above and rufous below, lacking the
elongated tail feathers and also any crimson in the plumage.
Only the forecrown is blue, with black above the bill. Neither
parrotfinch has been found in Bali.

# Checklist

THIS is a nominal checklist of the birds of mainland Java and Bali. It is not definitive. Wandering oceanic birds, rare migrants, and birds found only on some off-shore islands are not included. The names and taxonomy of some groups—notably the swiftlets—still remains confused, and bracketed alternative names are given where appropriate. Some rarer species, especially those of lowland forests, are known to be endangered, while the continued existence of some others requires confirmation. One endemic bird is already extinct, the Javan Wattled Lapwing.

Prefixes are used as follows:
*    Described in this book
E    Endemic
N    Migrant from north
S    Migrant from south (or east)
B    Bali (not recorded from Java)
J    Java (not recorded from Bali)
Migratory status (N or S) is not indicated where both resident and migratory forms occur.

GREBES

| | |
|---|---|
| Red-throated Little Grebe (Dabchick) | *Tachybaptus (Podiceps) ruficollis* |
| Black-throated Little Grebe | *Tachybaptus (Podiceps) novaehollandiae* |

TROPIC-BIRDS

| | |
|---|---|
| White-tailed Tropic-Bird | *Phaeton lepturus* |

## PELICANS

| | | |
|---|---|---|
| N | Spot-billed Pelican | *Pelecanus philippensis* |
| S | Australian Pelican | *Pelecanus conspicillatus* |

## BOOBIES

| | | |
|---|---|---|
| | Brown Booby | *Sula leucogaster* |

## CORMORANTS

| | | |
|---|---|---|
| * | Little Cormorant | *Phalacrocorax niger* |
| * | Little Black Cormorant | *Phalacrocorax sulcirostris* |
| B | Little Pied Cormorant | *Phalacrocorax melanoleucos* |
| * | Oriental Darter | *Anhinga melanogaster* |

## FRIGATE-BIRDS

| | | |
|---|---|---|
| * | Lesser Frigate-Bird | *Fregata ariel* |
| | Great Frigate-Bird | *Fregata minor* |

## HERONS, EGRETS, and BITTERNS

| | | |
|---|---|---|
| | Great-billed Heron | *Ardea sumatrana* |
| * | Grey Heron | *Ardea cinerea* |
| * | Purple Heron | *Ardea purpurea* |
| SB | White-faced Heron | *Ardea novaehollandiae* |
| | Little (Little Green) Heron | *Butorides striatus* |
| * | Black-crowned Night Heron | *Nycticorax nycticorax* |
| | Rufous Night Heron | *Nycticorax caledonicus* |
| * | Great Egret | *Egretta alba* |
| * | Plumed (Intermediate or Short-billed) Egret | *Egretta intermedia* |
| * | Little Egret | *Egretta garzetta* |
| * | Pacific Reef Egret | *Egretta sacra* |
| * | Cattle Egret | *Bubulcus ibis* |
| * | Javan Pond Heron | *Ardeola speciosa* |
| * | Cinnamon Bittern | *Ixobrychus cinnamomeus* |
| N | Yellow Bittern | *Ixobrychus sinensis* |
| | Black Bittern | *Ixobrychus (Dupetor) flavicollis* |

## STORKS

| | | |
|---|---|---|
| * | Milky Stork | *Mycteria (Ibis) cinerea* |
| * | Woolly-necked (White-necked) Stork | *Ciconia episcopus* |
| * | Lesser Adjutant | *Leptoptilos javanicus* |

## IBISES

| | | |
|---|---|---|
| * | Glossy Ibis | *Plegadis falcinellus* |
| * | Black-headed Ibis | *Threskiornis melanocephalus* |

## DUCKS

| | | |
|---|---|---|
| *J | Lesser Treeduck (Whistling Teal) | *Dendrocygna javanica* |
| * | Wandering Treeduck (Whistling Teal) | *Dendrocygna arcuata* |
| * | Grey Teal | *Anas gibberifrons* |
| NJ | Garganey | *Anas querqueduia* |
| | Pacific Black (or Australian Grey) Duck | *Anas superciliosa* |
| | White Pygmy Goose (Cotton Teal) | *Nettapus coromandelianus* |

## RAPTORS (including OSPREYS, HAWKS, EAGLES, FALCONS)

| | | |
|---|---|---|
| | Osprey | *Pandion haliaetus* |
| * | Honey Buzzard | *Pernis apivorus* |
| * | Black-shouldered Kite | *Elanus caeruleus* |
| * | Brahminy Kite | *Haliastur indus* |
| * | White-bellied Sea Eagle | *Haliaeetus leucogaster* |
| | Grey-headed (Larger) Fish-Eagle | *Ichthyophaga ichthyaetus* |
| * | Crested Serpent-Eagle | *Spilornis cheela* |
| *N | Japanese Sparrowhawk | *Accipiter gularis* |
| | Besra Sparrowhawk | *Accipiter virgatus* |
| N | Chinese Goshawk | *Accipiter soloensis* |
| | Crested Goshawk | *Accipiter trivirgatus* |
| | Rufous-winged Buzzard | *Butastur liventer* |
| * | Black Eagle | *Ictinaetus malayensis* |
| | Rufous-bellied Eagle | *Hieraaetus kieneri* |
| * | Changeable Hawk-Eagle | *Spizaetus cirrhatus* |
| *EJ | Javan Hawk-Eagle | *Spizaetus bartelsi* |
| | Black-thighed Falconet | *Microhierax fringillarius* |

| * | Spotted Kestrel | *Falco moluccensis* |
| | Oriental Hobby | *Falco severus* |
| | Peregrine Falcon | *Falco peregrinus* |

## GAME-BIRDS (including PHEASANTS and BUTTON-QUAILS)

| * | Blue-breasted Quail | *Coturnix chinensis* |
| *EJ | Javan (Chestnut-bellied) Partridge | *Arborophila javanica* |
| *J | Bar-backed (Grey-bellied) Partridge | *Arborophila orientalis* |
| * | Red Junglefowl | *Gallus gallus* |
| * | Green Junglefowl | *Gallus varius* |
| *J | Green Peafowl | *Pavo muticus* |
| * | Barred Button-Quail | *Turnix suscitator* |
| * | Little Button-Quail | *Turnix sylvatica* |

## CRAKES and RAILS

| * | Slaty-breasted Rail | *Rallus striatus* |
| | Red-legged (Malay Banded) Crake | *Rallina fasciata* |
| NJ | Baillon's Crake | *Porzana pusilla* |
| * | Ruddy (Ruddy-breasted) Crake | *Porzana fusca* |
| * | White-browed Crake | *Porzana cinerea* |
| * | White-breasted Waterhen | *Amaurornis phoenicurus* |
| N | Watercock | *Gallicrex cinerea* |
| * | Common Moorhen | *Gallinula chloropus* |
| * | Purple Swamphen (Coot) | *Porphyrio porphyrio* |
| | Common Coot | *Fulica atra* |

## JACANAS

| N | Pheasant-tailed Jacana | *Hydrophasianus chirurgus* |
| J | Bronze-winged Jacana | *Metopidius indicus* |

## PAINTED SNIPE

| J | Painted (Greater Painted) Snipe | *Rostratula benghalensis* |

## PLOVERS

| *EJ | Javan Wattled Lapwing (extinct) | *Vanellus macropterus* |

| | | |
|---|---|---|
| *N | (Lesser or Pacific) Golden Plover | *Pluvialis dominica* |
| N | Grey Plover | *Pluvialis squatarola* |
| *N | Little Ringed Plover | *Charadrius dubius* |
| *N | Kentish Plover | *Charadrius alexandrinus* |
| *E | Javan Sand Plover | *Charadrius javanicus* |
| * | Malay Sand Plover | *Charadrius peronii* |
| N | Greater Sand Plover | *Charadrius leschenaultii* |
| N | Lesser (Mongolian) Sand Plover | *Charadrius mongolus* |
| N | Oriental Plover | *Charadrius veredus* |

## WADERS (including SANDPIPERS, etc.)

| | | |
|---|---|---|
| *N | Eurasian Curlew | *Numenius arquata* |
| N | (Far) Eastern Curlew | *Numenius madagascariensis* |
| *N | Whimbrel | *Numenius phaeopus* |
| N | Black-tailed Godwit | *Limosa limosa* |
| N | Bar-tailed Godwit | *Limosa lapponica* |
| N | (Common) Redshank | *Tringa totanus* |
| N | Marsh Sandpiper | *Tringa stagnatilis* |
| N | (Common) Greenshank | *Tringa nebularia* |
| *N | Wood Sandpiper | *Tringa glareola* |
| N | Terek Sandpiper | *Xenus cinereus* |
| *N | Common Sandpiper | *Actitis hypoleucos* |
| N | Grey-tailed Tattler | *Heteroscellus brevipes* |
| N | Ruddy Turnstone | *Arenaria interpres* |
| N | Asian Dowitcher | *Limnodromus semipalmatus* |
| N | Pintail Snipe | *Gallinago stenura* |
| N | Swinhoe's Snipe | *Gallinago megala* |
| NJ | Common Snipe | *Gallinago gallinago* |
| | Indonesian (Dusky) Woodcock | *Scolopax saturara* |
| N | Great Knot | *Calidris tenuirostris* |
| N | Rufous-necked Stint | *Calidris ruficollis* |
| N | Long-toed Stint | *Calidris subminuta* |
| N | Sharp-tailed Sandpiper | *Calidris acuminata* |
| N | Curlew Sandpiper | *Calidris ferruginea* |
| N | Sanderling | *Calidris alba* |

STILTS
    White-headed
      (Black-winged) Stilt    *Himantopus leucocephalus*

PHALAROPES
  N    Red-necked Phalarope    *Phalaropus lobatus*

THICK-KNEES
    Great (Beach)
      Thick-knee    *Esacus magnirostris*

PRATINCOLES
  N    Oriental Pratincole    *Glareola maldivarum*
  S    Long-legged Pratincole    *Stiltia isabella*

TERNS
  *SN    Whiskered Tern    *Chlidonias hybridus*
  *N    White-winged (Black)
      Tern    *Chlidonias leucopterus*
  N    Gull-billed Tern    *Gelochelidon nilotica*
  *N    Common Tern    *Sterna hirundo*
    Roseate Tern    *Sterna dougallii*
    Black-naped Tern    *Sterna sumatrana*
  *    Bridled Tern    *Sterna anaethetus*
  *    Little Tern    *Sterna albifrons*
  *    Great Crested Tern    *Sterna bergii*
  N    Lesser Crested Tern    *Sterna bengalensis*

PIGEONS and DOVES
  J    Pin-tailed (Yellow-
      bellied) Green Pigeon    *Treron oxyura*
  J    Wedge-tailed Green
      Pigeon    *Treron sphenura*
  J    Large Green Pigeon    *Treron capellei*
  *    Grey-headed (-cheeked)
      Green Pigeon    *Treron griseicauda*
  J    Little Green Pigeon    *Treron olax*
  *    Pink-necked Green
      Pigeon    *Treron vernans*

| | | |
|---|---|---|
| J | Orange-breasted Green Pigeon | *Treron bicincta* |
| * | Pink-necked Fruit Dove | *Ptilinopus porphyreus* |
| * | Black-naped Fruit Dove | *Ptilinopus melanospila* |
| B | Black-backed (White-headed) Fruit Dove | *Ptilinopus cinctus* |
| * | Green Imperial Pigeon | *Ducula aenea* |
| | Pied Imperial Pigeon | *Ducula bicolor* |
| J | Mountain Imperial Pigeon | *Ducula badia* |
| * | Dark-backed (Black-backed) Imperial Pigeon | *Ducula lacernulata* |
| * | Little Cuckoo-Dove | *Macropygia ruficeps* |
| | Barred (Large) Cuckoo-Dove | *Macropygia unchall* |
| * | Brown (Sunda Red) Cuckoo-Dove | *Macropygia phasianella (emiliana)* |
| * | Spotted-necked Dove | *Streptopelia chinensis* |
| * | Javan Turtle-(Collared) Dove | *Streptopelia bitorquata* |
| * | Zebra (Peaceful, Barred Ground) Dove | *Geopelia striata* |
| | Green-winged Pigeon | *Chalcophaps indica* |

## PARROTS

| | | |
|---|---|---|
| * | Red-breasted (Moustached) Parakeet | *Psittacula alexandri* |
| *E | Javan (Yellow-throated) Hanging Parrot | *Loriculus pusillus* |

## CUCKOOS, MALKOHAS, and COUCALS

| | | |
|---|---|---|
| NJ | Large Hawk-Cuckoo | *Cuculus sparveriodes* |
| J | Lesser (Moustached) Hawk-Cuckoo | *Cuculus vagans* |
| J | Hodgson's Hawk-Cuckoo | *Cuculus fugax* |
| J | Indian Cuckoo | *Cuculus micropterus* |
| * | Oriental Cuckoo | *Cuculus saturatus* |
| *J | Banded Bay Cuckoo | *Cacomantis (Cuculus) sonneratii* |
| * | Plaintive Cuckoo | *Cacomantis (Cuculus) merulinus* |

| | | |
|---|---|---|
| * | Indonesian (Brush) Cuckoo | *Cacomantis sepulcralis (variolosus)* |
| J | Violet Cuckoo | *Chrysococcyx xanthorhynchus* |
| J | Malay (Little) Bronze Cuckoo | *Chrysococcyx minutillus* |
| S | Horsfield's Bronze Cuckoo | *Chrysococcyx basalis* |
| * | Drongo Cuckoo | *Surniculus lugubris* |
| * | (Common) Koel | *Eudynamys scolopacea* |
| * | Chestnut-breasted Malkoha | *Phaenicophaeus curvirostris* |
| *J | Red-billed Malkoha | *Phaenicophaeus javanicus* |
| * | Greater Coucal | *Centropus sinensis* |
| * | Lesser Coucal | *Centropus bengalensis* |
| *EJ | Sunda (Javan) Coucal | *Centropus nigrorufus* |

## OWLS

| | | |
|---|---|---|
| * | Barn Owl | *Tyto alba* |
| | Bay Owl | *Phodilus badius* |
| * | Collared Scops Owl | *Otus bakkamoena (lempiji)* |
| J | Reddish Scops Owl | *Otus rufescens* |
| *EJ | Javan Scops Owl | *Otus angelinae* |
| J | Rajah's Scops Owl | *Otus brookei* |
| * | Barred Eagle-Owl | *Bubo sumatranus* |
| * | Buffy Fish-Owl | *Ketupa (Bubo) ketupu* |
| *E | Javan Barred Owlet | *Glaucidium castanopterum* |
| | Brown Hawk-Owl | *Ninox scutulata* |
| J | Spotted Wood Owl | *Strix seloputo* |
| J | Brown Wood Owl | *Strix leptogrammica* |

## FROGMOUTHS

| | | |
|---|---|---|
| J | Javan Frogmouth | *Batrachostomus javensis* |

## NIGHTJARS

| | | |
|---|---|---|
| * | Large-tailed Nightjar | *Caprimulgus macrurus* |
| * | Savanna Nightjar | *Caprimulgus affinis* |
| NJ | Grey Nightjar | *Caprimulgus indicus* |
| J | Salvadori's Nightjar | *Caprimulgus pulchellus* |

## SWIFTS and TREESWIFTS

| | | |
|---|---|---|
| * | Linchi (White-bellied) Swiftlet | *Collocalia (Aerodramus) linchi* |

95

| | | |
|---|---|---|
| * | Edible-nest Swiftlet | *Collocalia (Aerodramus) fuciphaga* |
| *J | Black-nest Swiftlet | *Collocalia (Aerodramus) maxima* |
| | Himalayan Swiftlet | *Collocalia (Aerodramus) brevirostris* |
| | Mossy-nest Swiftlet | *Collocalia (Aerodramus) vanikorensis* |
| J | Giant Swiftlet | |
| | (Waterfall Swift) | *Collocalia (Hydrochous) gigas* |
| N | White-throated | |
| | Needletail | *Hirundapus caudacutus* |
| N | White-vented Needletail | *Hirundapus cochinchinensis* |
| | Brown Needletail | *Hirundapus giganteus* |
| J | Silver-rumped Spinetail | *Chaetura leucopygialis* |
| * | House Swift | *Apus affinis* |
| N | White-rumped | |
| | (Fork-tailed) Swift | *Apus pacificus* |
| * | (Asian) Palm Swift | *Cypsiurus balasiensis* |
| | Grey-rumped Treeswift | *Hemiprocne longipennis* |

## TROGONS

| | | |
|---|---|---|
| *J | Blue-tailed Trogon | *Harpactes reinwardtii* |
| *J | Orange-breasted Trogon | *Harpactes oreskios* |

## KINGFISHERS

| | | |
|---|---|---|
| N | Common Kingfisher | *Alcedo atthis* |
| * | Deep Blue (Blue-eared) | |
| | Kingfisher | *Alcedo meninting* |
| * | Small Blue Kingfisher | *Alcedo coerulescens* |
| J | Blue-banded Kingfisher | *Alcedo euryzona* |
| | Dwarf (Rufous-backed) | |
| | Kingfisher | *Ceyx rufidorsus (erithacus)* |
| * | Collared (White- | |
| | collared) Kingfisher | *Halcyon chloris* |
| *S | Sacred Kingfisher | *Halcyon sancta* |
| *E | Javan Kingfisher | *Halcyon cyaniventris* |
| J | Ruddy Kingfisher | *Halcyon coromanda* |
| NJ | Black-capped Kingfisher | *Halcyon pileata* |
| * | Stork-billed Kingfisher | *Pelargopsis (Halcyon) capensis* |
| J | Banded Kingfisher | *Lacedo pulchella* |

## BEE-EATERS

| | | |
|---|---|---|
| * | Chestnut-headed | |
| | Bee-eater | *Merops leschenaulti* |

| | | |
|---|---|---|
| *N | Brown-breasted (Blue-tailed) Bee-eater | *Merops philippinus* |
| J | Blue-throated Bee-eater | *Merops viridis* |
| SB | Rainbow Bee-eater | *Merops ornatus* |

## ROLLERS

| | | |
|---|---|---|
| | Dollarbird (Broad-billed Roller) | *Eurystomus orientalis* |

## HORNBILLS

| | | |
|---|---|---|
| * | Wreathed Hornbill | *Rhyticeros undulatus* |
| * | (Southern) Pied Hornbill | *Anthracoceros coronatus (convexus)* |
| *J | Rhinoceros Hornbill | *Buceros rhinoceros* |

## BARBETS

| | | |
|---|---|---|
| *E | Black-banded Barbet | *Megalaima javensis* |
| *E | Blue-crowned Barbet | *Megalaima armillaris* |
| *EJ | Brown-throated Barbet | *Megalaima corvina* |
| * | Blue-eared (Little) Barbet | *Megalaima australis* |
| * | Coppersmith Barbet | *Megalaima haemacephala* |
| * | Lineated Barbet | *Megalaima lineata* |

## WOODPECKERS

| | | |
|---|---|---|
| J | Rufous Piculet | *Sasia abnormis* |
| J | Rufous Woodpecker | *Celeus (Micropternus) brachyurus* |
| | Laced Woodpecker | *Picus vittatus* |
| *J | Crimson-winged Woodpecker | *Picus puniceus* |
| J | Checker-throated Woodpecker | *Picus mentalis* |
| J | Banded Woodpecker | *Picus miniaceus* |
| * | Common Golden-backed Woodpecker | *Dinopium javanense* |
| J | Buff-rumped Woodpecker | *Meiglyptes tristis* |
| J | Great Slaty Woodpecker | *Mulleripicus pulverulentus* |
| | White-bellied (Black) Woodpecker | *Dryocopus javensis* |
| * | Fulvous-breasted (Pygmy) Woodpecker | *Dendrocopus (Picoides) macei* |

97

|   |   |   |
|---|---|---|
|   | Brown-capped (Pygmy) Woodpecker | *Dendrocopus (Picoides) moluccensis* |
| J | Grey-and-Buff Woodpecker | *Hemicircus concretus* |
| *J | Orange-backed Woodpecker | *Reinwardtipicus (Chrysocolaptes) validus* |
|   | Greater Golden-backed Woodpecker | *Chrysocolaptes lucidus* |

## BROADBILLS

|   |   |   |
|---|---|---|
| J | Banded Broadbill | *Eurylaimus javanicus* |

## PITTAS

|   |   |   |
|---|---|---|
| * | Banded Pitta | *Pitta guajana* |
| J | Hooded Pitta | *Pitta sordida* |
| NJ | (Lesser) Blue-winged Pitta | *Pitta moluccensis* |

## LARKS

|   |   |   |
|---|---|---|
| * | Singing Bushlark | *Mirafra javanica* |

## SWALLOWS

|   |   |   |
|---|---|---|
| *N | Barn Swallow | *Hirundo rustica* |
| * | Pacific Swallow | *Hirundo tahitica* |
| * | Red-rumped (Larger Striated) Swallow | *Hirundo striolata* |
| N | Asian House Martin | *Delichon dasypus* |

## CUCKOO-SHRIKES, TRILLERS, and MINIVETS

|   |   |   |
|---|---|---|
| * | Black-winged Flycatcher-Shrike | *Hemipus hirundinaceus* |
| J | Large Wood Shrike | *Tephrodornis gularis* |
|   | Large (Black-masked) Cuckoo-Shrike | *Coracina novaehollandiae (javensis)* |
| *J | Black-faced Cuckoo-Shrike | *Coracina larvata* |
|   | Lesser Cuckoo-Shrike | *Coracina fimbriata* |
| *J | Pied Triller· | *Lalage nigra* |
| * | White-winged Triller | *Lalage sueuri* |
| *J | Sunda Minivet | *Pericrocotus miniatus* |

| | | |
|---|---|---|
| * | Scarlet Minivet | *Pericrocotus flammeus* |
| * | Small Minivet | *Pericrocotus cinnamomeus* |

## IORAS and LEAFBIRDS

| | | |
|---|---|---|
| * | Common Iora | *Aegithina tiphia* |
| | Greater Green Leafbird | *Chloropsis sonnerati* |
| *J | Blue-winged Leafbird | *Chloropsis cochinchinensis* |

## BULBULS

| | | |
|---|---|---|
| *J | Straw-headed Bulbul | *Pycnonotus zeylanicus* |
| * | Black-headed Bulbul | *Pycnonotus atriceps* |
| * | Ruby-throated (Black-crested) Bulbul | *Pycnonotus (melanicterus) dispar* |
| J | Scaly-breasted Bulbul | *Pycnonotus squamatus* |
| * | Sooty-headed (Golden-vented) Bulbul | *Pycnonotus aurigaster* |
| * | Yellow-vented Bulbul | *Pycnonotus goiavier* |
| * | Orange-spotted Bulbul | *Pycnonotus bimaculatus* |
| J | Olive-winged Bulbul | *Pycnonotus plumosus* |
| J | Cream-vented Bulbul | *Pycnonotus simplex* |
| | Grey-cheeked (Scrub) Bulbul | *Criniger bres* |
| J | Javan Streaked (Mountain) Bulbul | *Hypsipetes (Ixos) virescens* |

## DRONGOS

| | | |
|---|---|---|
| * | Black Drongo | *Dicrurus macrocercus* |
| * | Ashy Drongo | *Dicrurus leucophaeus* |
| * | Spangled (Hair-crested) Drongo | *Dicrurus hottentottus* |
| NJ | Crow-billed Drongo | *Dicrurus annectans* |
| *J | Lesser Racket-tailed Drongo | *Dicrurus remifer* |
| * | Greater Racket-tailed Drongo | *Dicrurus paradiseus* |

## ORIOLES

| | | |
|---|---|---|
| * | Black-naped Oriole | *Oriolus chinensis* |
| J | Dark-throated (Black-headed) Oriole | *Oriolus xanthonotus* |

| *J | Black-and-Crimson | |
| | Oriole | *Oriolus cruentus* |
| J | Fairy Bluebird | *Irena puella* |

## CROWS

| * | Large-billed Crow | *Corvus macrorhynchos* |
| * | Slender-billed Crow | *Corvus enca* |
| J | Crested Jay | *Platylophus galericulatus* |
| * | Racket-tailed Treepie | *Crypsirina temia* |
| *J | Short-tailed Green | |
| | Magpie | *Cissa thalassina* |

## TITS

| * | Great Tit | *Parus major* |
| *EJ | Pygmy Tit | *Psaltria exilis* |

## NUTHATCHES

| *J | Blue Nuthatch | *Sitta azurea* |
| *J | Velvet-fronted Nuthatch | *Sitta frontalis* |

## BABBLERS

| *J | Black-capped Babbler | *Pellorneum capistratum* |
| J | Temminck's Babbler | *Trichastoma tickelli (pyrrogenys)* |
| * | Horsfield's Babbler | *Trichastoma sepiarium* |
| J | Scaly-crowned Babbler | *Malacopteron cinereum* |
| * | Chestnut-backed | |
| | Scimitar Babbler | *Pomatorhinus montanus* |
| J | Large Wren-Warbler | *Napothera macrodactyla* |
| J | Eye-browed (Small) | |
| | Wren-Babbler | *Napothera epilepidota* |
| *J | Pygmy Wren-Babbler | *Pnoepyga pusilla* |
| *EJ | White-breasted Babbler | *Stachyris grammiceps* |
| *EJ | White-collared Babbler | *Stachyris thoracica* |
| *E | Pearl-cheeked Babbler | *Stachyris melanothorax* |
| *J | Striped Tit-Babbler | *Macronous gularis* |
| *EJ | Javan (Grey-faced, | |
| | Yellow-collared) | |
| | Tit-Babbler | *Macronous flavicollis* |
| J | Chestnut-capped Babbler | *Timalia pileata* |
| *EJ | Red-fronted Laughing- | |
| | Thrush | *Garrulax rufifrons* |

| *J | White-browed Shrike-Babbler | *Pteruthius flaviscapis* |
| *J | Chestnut-fronted Shrike-Babbler | *Pteruthius aenobarbus* |
| *EJ | Javan Fulvetta | *Alcippe pyrrhoptera* |
| *EJ | Spotted Crocias | *Crocias albonotatus* |

## THRUSHES, ROBINS, CHATS, etc.

| * | Lesser Shortwing | *Brachypteryx leucophrys* |
| *J | White-browed Shortwing | *Brachypteryx montana* |
| * | Magpie Robin | *Copsychus saularis* |
| *J | White-rumped Shama | *Copsychus malabaricus* |
| * | White-crowned Forktail | *Enicurus leschenaulti* |
| *J | Lesser Forktail | *Enicurus velatus* |
| J | Sunda Blue Robin | *Cinclidium diana* |
| * | Pied Bushchat | *Saxicola caprata* |
| *EJ | Javan Cochoa | *Cochoa azurea* |
| * | Sunda Whistling Thrush | *Myophoneus glaucinus* |
| *J | Blue Whistling Thrush | *Myophoneus caeruleus* |
| | Chestnut-capped Thrush | *Zoothera interpres* |
| * | Orange-headed Thrush | *Zoothera citrina* |
| | Sunda (Ground) Thrush | *Zoothera andromedae* |
| N | Siberian Thrush | *Zoothera sibirica* |
| | Scaly (White's) Thrush | *Zoothera dauma* |
| J | Island Thrush | *Turdus poliocephalus* |
| NJ | Eye-browed Thrush | *Turdus obscurus* |

## WARBLERS

| * | Fly-eater | *Gerygone sulphurea* |
| * | Sunda (Flycatcher-) Warbler | *Seicercus grammiceps* |
| | Yellow-bellied (Flycatcher-) Warbler | *Abroscopus superciliaris* |
| N | Arctic Leaf Warbler | *Phylloscopus borealis* |
| NJ | Eastern Crowned Leaf Warbler | *Phylloscopus coronatus* |
| * | Mountain Leaf Warbler | *Phylloscopus trivirgatus* |
| J | Clamorous Reed Warbler | *Acrocephalus stentoreus* |

| | | |
|---|---|---|
| NJ | Great (Eastern) Reed Warbler | *Acrocephalus arundinaceus* |
| NJ | Pallas' (Grasshopper-) Warbler | *Locustella certhiola* |
| NJ | Lanceolated Warbler | *Locustella lanceolata* |
| * | Striated (Grass) Warbler | *Megalurus palustris* |
| *J | Long-tailed Tailorbird | *Orthotomus sutorius* |
| * | Ashy Tailorbird | *Orthotomus ruficeps (sepium)* |
| * | Mountain Tailorbird | *Orthotomus cuculatus* |
| * | Bar-winged Prinia (Wren-Warbler) | *Prinia familiaris* |
| J | Yellow-bellied Prinia (Wren-Warbler) | *Prinia flaviventris* |
| J | Tawny-flanked Prinia (Wren-Warbler) | *Prinia subflava (inornata)* |
| J | Brown Prinia (Wren-Warbler) | *Prinia polychroa* |
| | Zitting Cisticola (Fantail Warbler) | *Cisticola juncidis* |
| | Bright-capped Cisticola (Fantail Warbler) | *Cisticola exilis* |
| *EJ | Javan Tesia (Ground Warbler) | *Tesia superciliaris* |
| | Mueller's (Mountain) Bush Warbler | *Cettia vulcania* |
| *J | Javan Bush (Scrub) Warbler | *Bradypterus montis (seebohmi)* |

## FLYCATCHERS and WHISTLERS

| | | |
|---|---|---|
| | Fulvous-chested (Olive-backed) Flycatcher | *Rhinomyias olivacea* |
| NJ | Sooty (Dark-sided) Flycatcher | *Muscicapa sibirica* |
| NJ | Asian Brown Flycatcher | *Muscicapa latirostris* |
| NJ | Ferruginous Flycatcher | *Muscicapa ferruginea* |
| *J | Indigo Flycatcher | *Muscicapa indigo* |
| NJ | Yellow-rumped Flycatcher | *Ficedula zanthopygia* |
| NJ | Mugimaki Flycatcher | *Ficedula mugimaki* |

| | | |
|---|---|---|
| * | Snowy-browed Flycatcher | *Ficedula hyperythra* |
| J | Rufous-chested Flycatcher | *Ficedula dumetoria* |
| | Little Pied Flycatcher | *Ficedula westermanni* |
| NJ | Blue-and-White Flycatcher | *Cyanoptila cyanomelana* |
| J | Pale Blue Flycatcher | *Cyornis unicolor* |
| *J | Hill Blue Flycatcher | *Cyornis banyumas* |
| J | Mangrove Blue Flycatcher | *Cyornis rufigastra* |
| | Grey-headed Flycatcher | *Culicapa ceylonensis* |
| * | Pied Fantail | *Rhipidura javanica* |
| *EJ | Red-tailed Fantail | *Rhipidura phoenicura* |
| *EJ | White-bellied Fantail | *Rhipidura euryura* |
| | Black-naped Monarch | *Hypothymis azurea* |
| J | Maroon-breasted Monarch | *Philentoma velatum* |
| * | Asian Paradise Flycatcher | *Terpsiphone paradisi* |
| | Mangrove Whistler | *Pachycephala cinerea (grisola)* |
| | Golden Whistler | *Pachycephala pectoralis* |

## WAGTAILS and PIPITS

| | | |
|---|---|---|
| *N | Grey Wagtail | *Motacilla cinerea* |
| *N | Yellow Wagtail | *Motacilla flava* |
| NJ | Forest Wagtail | *Dendronanthus indicus* |
| * | Richard's Pipit | *Anthus novaeseelandiae* |

## WOOD-SWALLOWS

| | | |
|---|---|---|
| * | White-breasted Wood-Swallow (Swallow-Shrike) | *Artamus leucorhynchus* |

## SHRIKES

| | | |
|---|---|---|
| * | Long-tailed (Rufous-backed) Shrike | *Lanius schach* |
| N | Brown Shrike | *Lanius cristatus* |
| N | Tiger (Thick-billed) Shrike | *Lanius tigrinus* |

## STARLINGS and MYNAS

| | | |
|---|---|---|
| * | Philippine Glossy Starling | *Aplonis panayensis* |
| | Lesser (Short-tailed) Glossy Starling | *Aplonis minor* |
| NJ | Daurian (Purple-backed) Starling | *Sturnus sturninus* |
| * | Asian Pied Starling | *Sturnus contra* |
| *E | Black-winged Starling | *Sturnus melanopterus* |
| * | White-vented Myna | *Acridotheres javanicus* |
| *EB | Bali White (Rothschild's) Myna | *Leucopsar rothschildi* |
| * | Grackle or Hill (Talking) Myna | *Gracula religiosa* |

## HONEYEATERS

| | | |
|---|---|---|
| B | Brown Honeyeater | *Lichmera indistincta* |

## SUNBIRDS and SPIDERHUNTERS

| | | |
|---|---|---|
| * | Brown-throated Sunbird | *Anthreptes malacensis* |
| *J | Ruby-cheeked Sunbird | *Anthreptes singalensis* |
| * | Olive-backed (Yellow-bellied) Sunbird | *Nectarinia jugularis* |
| J | Purple-throated (Van Hasselt's) Sunbird | *Nectarinia sperata* |
| J | Copper-throated (Macklot's) Sunbird | *Nectarinia calcostetha* |
| *EJ | Kuhl's Sunbird | *Aethopyga eximia* |
| *EJ | Scarlet Sunbird | *Aethopyga mystacalis* |
| J | Crimson Sunbird | *Aethopyga siparaja* |
| * | Little Spiderhunter | *Arachnothera longirostra* |
| J | Long-billed Spiderhunter | *Arachnothera robusta* |
| J | Yellow-eared Spiderhunter | *Arachnothera chrysogenys* |
| | Grey-breasted Spiderhunter | *Arachnothera affinis* |

## FLOWERPECKERS

| | | |
|---|---|---|
| J | Crimson-breasted Flowerpecker | *Prionochilus percussus* |

| | | |
|---|---|---|
| * | Scarlet-headed<br>Flowerpecker | *Dicaeum trochileum* |
| J | Thick-billed<br>Flowerpecker | *Dicaeum agile* |
| J | Yellow-vented<br>Flowerpecker | *Dicaeum chrysorrheum* |
| | Plain Flowerpecker | *Dicaeum concolor* |
| * | Orange-bellied<br>Flowerpecker | *Dicaeum trigonostigma* |
| *E | Javan Fire-breasted<br>Flowerpecker | *Dicaeum sanguinolentum* |

## WHITE-EYES

| | | |
|---|---|---|
| * | Oriental White-eye | *Zosterops palpebrosus* |
| J | Javan White-eye | *Zosterops flavus* |
| | Lemon-bellied<br>(Mangrove)<br>White-eye | *Zosterops chloris* |
| | Mountain White-eye | *Zosterops montanus* |
| *E | Javan Grey-fronted<br>White-eye | *Lophozosterops javanicus* |

## SPARROWS, WEAVERS, and MUNIAS

| | | |
|---|---|---|
| * | (Eurasian) Tree Sparrow | *Passer montanus* |
| * | Streaked Weaver | *Ploceus manyar* |
| * | Baya Weaver | *Ploceus philippinus* |
| *J | Golden Weaver | *Ploceus hypoxanthus* |
| | Red Avadavat | *Amandava amandava* |
| *(E) | Java Sparrow | *Padda oryzivora* |
| * | White-headed Munia | *Lonchura maja* |
| | Chestnut Munia | *Lonchura malacca* |
| * | Scaly-breasted Munia | *Lonchura punctulata* |
| * | Javan White-bellied<br>Munia | *Lonchura leucogastroides* |
| *J | Pin-tailed Parrotfinch | *Erythrura prasina* |
| *J | Tawny-breasted<br>(Bamboo) Parrotfinch | *Erythrura hyperythra* |

## FINCHES

| | | |
|---|---|---|
| J | Sunda Serin<br>(Malay Goldfinch) | *Serinus estherae* |

# Index to Genera, Systematic Section

*References in italics refer to colour plate numbers*

*Accipiter*, 11, *3*
*Acridotheres*, 78
*Actitis*, 19, *5*
Adjutant, 7
*Aegithina*, 53, *14*
*Aethopyga*, 81, *23*
*Alcedo*, 38, *10*
*Alcippe*, 65
*Amaurornis*, 15, *4*
*Anas*, 8–9
*Anhinga*, 2
*Anthracoceros*, 42
*Anthreptes*, 80, *23*
*Anthus*, 76
*Aplonis*, 77, *22*
*Apus*, 35, *9*
*Arachnothera*, 81, *23*
*Arborophila*, 14
*Ardea*, 3, *1*
*Ardeola*, 4, *1*
*Artamus*, 76, *22*

Babblers, 61, *18*
Babblers, Scimitar, 62, *18*
Babblers, Shrike-, 64, *18*
Babblers, Tit-, 64
Babblers, Wren-, 62, *18*
Barbets, 43, *12*
Bee-eaters, 40, *11*
Bittern, 6, *1*
*Brachypteryx*, 66, *19*
*Bradypterus*, 72
*Bubo*, 32
*Bubulcus*, 4, *1*
*Buceros*, 42
Bulbuls, 54, *15*

Bush Warbler, 72
Bushchat, 68, *19*
Bushlark, 48, *13*
Button-Quail, 13
Buzzard, 12

*Cacatua*, 26
*Cacomantis*, 28, *7*
*Caprimulgus*, 34, *9*
*Centropus*, 30, *8*
*Charadrius*, 18, *5*
Chat, 68, *19*; *see* Bushchat
*Chlidonias*, 21
*Chloropsis*, 53
*Ciconia*, 7, *2*
*Cissa*, 59, *17*
Cochoa, 69
*Cochoa*, 69
Cockatoo, 26
*Collocalia*, 36, *9*
*Copsychus*, 66, *19*
*Coracina*, 51
Cormorants, 1
*Corvus*, 58, *16*
*Coturnix*, 12, *4*
Coucals, 30, *8*
Crakes, 16, *4*
Crocias, 65
*Crocias*, 65
Crows, 58, *16*
*Crypsirina*, 59, *17*
Cuckoo-Dove, 24, *6*; *see* Doves
Cuckoo-Shrike, 51
Cuckoos, 28, *7*
*Cuculus*, 29
Curlew, 19

*Cyornis*, 73
*Cypsiurus*, 36, *9*

Darter, 2
*Dendrocopus*, 47, *13*
*Dendrocygna*, 8, *2*
*Dicaeum*, 82, *23*
*Dicrurus*, 56, *16*
*Dinopium*, 46
Doves, Fruit, 24, *6*
Doves, Cuckoo-, 24, *6*
Doves, Turtle-, 25, *7*
Drongos, 56, *16*
Duck, 8, *2*; *see* Treeduck
*Ducula*, 23–4

Eagle-Owl, 32; *see* Owl
Eagles, 10, *3*
Egrets, 4, *1*
*Egretta*, 4
*Elanus*, 9, *3*
*Enicurus*, 67, *19*
*Erythrura*, 87, *24*
*Eudynamys*, 30

*Falco*, 12, *3*
Fantails, 74, *21*
*Ficedula*, 74, *21*
Finch, Parrot, 87, *24*; *see* Parrotfinch
Fish-Owl, 32; *see* Owl
Flowerpeckers, 82, *23*
Flycatcher-Shrike, 51, *14*
Flycatchers, 73, *21*
Fly-eater, 70, *20*
Forktails, 67, *19*
*Fregata*, 3
Frigate-Birds, 3
Fruit Dove, 24, *6*; *see* Doves
Fulvetta, 65

*Gallinula*, 16
*Gallus*, 13, *4*
*Garrulax*, 64, *18*
*Geopelia*, 26, *7*

*Gerygone*, 70, *20*
*Glaucidium*, 33
Grackle, 78
*Gracula*, 78
Green Pigeons, 23, *6*; *see* Pigeons

*Halcyon*, 39, *10*
*Haliaeetus*, 10, *3*
*Haliastur*, 10
Hanging Parrot, 27, *7*; *see* Parrot
*Harpactes*, 37, *11*
Hawk-Eagles, 11; *see* Eagles
Hawk, Sparrow, 11, *3*; *see* Sparrowhawk
*Hemipus*, 51, *14*
Heron, Night, 5, *1*
Heron, Pond, 4, *1*
Herons, 3, *1*
*Hirundo*, 49, *13*
Hornbills, 42, *11*

Ibises, 8, *2*
*Ictinaetus*, 11
Imperial Pigeon, 23; *see* Pigeons
Iora, 53, *14*
*Ixobrychus*, 6, *1*

Junglefowl, 13, *4*

Kestrel, 12, *3*
*Ketupa*, 32
Kingfishers, 38, *10*
Kites, 9, *3*
Koel, 30

*Lalage*, 50, *14*
*Lanius*, 77, *22*
Lapwing, 18, *frontispiece*
Lark, 48, *13*; *see* Bushlark
Laughing-Thrush, 64, *18*
Leafbird, 53
Leaf Warbler, 70
*Leptoptilos*, 7
*Leucopsar*, 79
*Lonchura*, 86, *24*

*Lophozosterops*, 83
*Loriculus*, 27, *7*

*Macronous*, 64
*Macropygia*, 24, *6*
Magpie, Green, 59, *17*
Magpie Robin, 66, *19*
Malkoha, 30, *8*
*Megalaima*, 43, *12*
*Megalurus*, 71, *20*
*Merops*, 40, *11*
Minivets, 52, *14*
*Mirafra*, 48, *13*
Moorhen, 16
*Motacilla*, 75, *22*
Munias, 86, *24*
*Muscicapa*, 73, *21*
*Mycteria*, 7
Mynas, 78
*Myophoneus*, 68, *19*

*Nectarinia*, 80
Night Heron, 5, *1*
Nightjars, 34, *9*
*Numenius*, 19, *5*
Nuthatches, 60, *17*
*Nycticorax*, 5, *1*

Orioles, 57, *16*
*Oriolus*, 57, *16*
*Orthotomus*, 71, *20*
*Otus*, 33, *8*
Owlet, 33
Owls, 32, *8*

*Padda*, 85, *24*
Parakeet, 27, *7*
Parrot, Hanging, 27, *7*
Parrotfinches, 87, *24*
Partridges, 14
*Parus*, 60, *17*
*Passer*, 84, *24*
*Pavo*, 14
Peafowl, 14

*Pelargopsis*, 39
*Pellorneum*, 62
*Pericrocotus*, 52, *14*
*Pernis*, 12
*Phaenicophaeus*, 30, *8*
*Phalacrocorax*, 1
*Phylloscopus*, 70
*Picus*, 46, *13*
Pigeons, Green, 23, *6*
Pigeons, Imperial, 23–4
Pipit, 76
Pitta, 47, *13*
*Pitta*, 47, *13*
*Plegadis*, 8, *2*
*Ploceus*, 85, *24*
Plovers, 17, 5
*Pluvialis*, 17, 5
*Pnoepyga*, 62, *18*
*Pomatorhinus*, 62, *18*
Pond Heron, 4, *1*
*Porphyrio*, 15, *4*
*Porzana*, 16, *4*
Prinia, 72, *20*
*Prinia*, 72, *20*
*Psaltria*, 60
*Psittacula*, 27, *7*
*Pteruthius*, 64, *18*
*Ptilinopus*, 24, *6*
*Pycnonotus*, 54, *15*

Quails, 12, *4*

Rail, 16
*Rallus*, 16
*Reinwardtipicus*, 46
*Rhipidura*, 74, *21*
*Rhyticeros*, 42, *11*
Robin, Magpie, 66, *19*

Sandpipers, 19, 5
*Saxicola*, 68, *19*
Scimitar Babbler, 62, *18*; *see* Babblers
Scops Owl, 33, *8*; *see* Owls
Sea Eagle, 10, *3*

*Seicercus*, 70, *20*
Serpent-Eagle, 10, *3*
Shama, 67
Shortwings, 66, *19*
Shrike, 77, *22*
Shrike, Cuckoo-, 51
Shrike, Flycatcher-, 51, *14*
Shrike-Babblers, 64, *18*
*Sitta*, 60, *17*
Sparrowhawk, 11, *3*
Sparrows, 84, *24*
Spiderhunter, 81, *23*
*Spilornis*, 10, *3*
*Spizaetus*, 11
*Stachyris*, 63, *18*
Starlings, 77, *22*
*Sterna*, 20
Storks, 7, *2*
*Streptopelia*, 25, *6*, *7*
*Sturnus*, 78, *22*
Sunbirds, 80, *23*
*Surniculus*, 29
Swallows, 49, *13*
Swallows, Wood-, 76, *22*
Swamphen, 15, *4*
Swiftlets, 36, *9*
Swifts, 35, *9*

Tailorbirds, 71, *20*
Teal, 8–9
Terns, 20
*Terpsiphone*, 75, *21*
Tesia, 72
*Tesia*, 72

*Threskiornis*, 8
Thrush, 69, *19*
Thrush, Laughing-, 64, *18*
Thrush, Whistling, 68, *19*
Tit-Babbler, 64; *see* Babblers
Tits, 60, *17*
Treeducks, 8, *2*
Treepie, 59, *17*
*Treron*, 23, *6*
*Trichastoma*, 61, *18*
Trillers, 50, *14*
*Tringa*, 20
Trogons, 37, *11*
*Turnix*, 13
Turtle-Dove, 25, *7*; *see* Doves
*Tyto*, 32, *8*

*Vanellus*, 18, *frontispiece*

Wagtails, 75, *22*
Warblers, 70, *20*
Waterhen, 15, *4*
Weavers, 85, *24*
Whimbrel, 19, *5*
Whistling Teal, 8, *2*; *see* Treeducks
Whistling Thrush, 68, *19*; *see* Thrush
White-eyes, 83, *23*
Wren-Babbler, 62, *18*        ·
Wren-Warbler, 72, *20*; *see* Prinia
Woodpeckers, 46, *13*
Wood-Swallow, 76, *22*

*Zosterops*, 83, *23*
*Zoothera*, 69, *19*